Emerging Technologies in Agriculture, Livestock, and Climate

Abid Yahya

Emerging Technologies in Agriculture, Livestock, and Climate

 Springer

Abid Yahya
Botswana International University of Science and Technology
Palapye, Botswana

ISBN 978-3-030-33489-5   ISBN 978-3-030-33487-1   (eBook)
https://doi.org/10.1007/978-3-030-33487-1

This Springer imprint is published by the registered company Springer Nature Switzerland AG
The registered company address is: Gewerbestrasse 11, 6330 Cham, Switzerland

Dedicated to my family for their love, support, and sacrifice along the path of my academic pursuits, especially to my late father, who took me to school.

Abid Yahya

Preface

The impossible has become achievable in Africa due to the fastest growth in the mobile telecommunication industry. Africa, the second-biggest mobile market, has sparked some digital innovations, with diverse applications in the sector of education, health, ICT, agriculture, governance, finance, energy, and tourism.

The rapid growth of tech hubs in Africa gives birth to digital entrepreneurship ecosystem by networking entrepreneurs, designers, developers, and prospective investors.

This book consists of three (3) chapters and is organized as follows.

Chapter 1 provides the Internet of Things (IoTs) based Smart Agriculture System. The system aims at improving agricultural production in Botswana by remotely monitoring farms of all types using the IoTs. The system uses sensors to monitor different parameters for these living things to ensure that a proper environment is maintained at all times. This chapter gives a detailed explanation of the data loggers which transmit data wirelessly to a central point located on the farm. The central location is referred to as a gateway, and it is where the farm employees can visualize and analyze data. The on-farm network of sensors and gateway is linked to an online server through General Packet Radio Service (GPRS) and Satellite to allow for remote online data acquisition (DAQ).

In this chapter, low cost and reliable wireless data acquisition systems are implemented in real time in a banana field. The moisture stress, reducing the usage of excessive water, rapid growth of the weeds is achieved with the implementation of sensor-based site-specific irrigation. IoT-based remote control of irrigation can also be achieved in the system. The implemented system can be used to transfer the fertilizer and other chemicals to the field with the help of adding new sensors and valves.

Chapter 2 presents a system to detect foot and mouth disease as early as possible in a herd of cattle using wireless sensor networks. The system combines animal behavior and sensor values to determine the status of the cattle in terms of foot and mouth disease.

The system combines animal behavior and sensor values to determine the status of the cattle in terms of foot and mouth disease. The system first measures an average response of a cow under normal circumstances with the focus being on the measurements of body temperature, distance covered, and feeding rate every two hours. Data regarding the state of the cow is sent at certain time intervals to the farmers using ZigBee via a gateway.

This chapter provides results of the trials performed on the project. A signal is acquired from a cow sensor node and transmitted to the gateway which is interfaced to a computer running LABVIEW software as the graphical user interface of the system. The trials were done both for negatively and positively diagnosed cows.

Chapter 3 gives an overview of environmental monitoring systems using wireless sensor network, big data, and IoTs. This chapter outlines the effect of climate change on wild animals and also discusses innovations in response to climate change.

Electric fences are commonly used to control and manage the movement of animals in game reserves, private game and farms to restrict intruders such as unwanted predators and humans from the bound area. Wireless sensor network-based system for intruder detection and monitoring is presented in this system to minimize the human-animal disputes in Africa.

It is challenging to monitor elephants movement since these huge animals travel for very long distances. The biggest challenge in the existing wireless anti-poaching system is the limited network or no coverage. As a result, the non-monitored animals are simply subjected to poaching. Taking advantage of WSN, an anti-poaching system is proposed in this chapter.

Palapye, Botswana   Abid Yahya
Palapye, Botswana   Joseph M. Chuma
Palapye, Botswana   Ravi Samikannu
Palapye, Botswana   Adamu Murtala Zungeru
Palapye, Botswana   Caspar K. Lebekwe
Palapye, Botswana   Taolo Tlale
Palapye, Botswana   Leatile Marata

Acknowledgments

Thank you to Karabo D. Bogaisang, Odirile G. Gamoshe, Dennis M. Maina, and Tshiamo Tsheole without whose contributions and support this book would not have been written.

The authors would like to express their special gratitude and thanks to Botswana International University of Science and Technology (BIUST); HRDC Botswana; Karakoram International University (KIU), Gilgit, Pakistan; University of Peradeniya, Sri Lanka; Sarhad University of Science & Information Technology (SUIT), Pakistan; City University of Science & Information Technology (CUSIT), Pakistan; and Muni University, Uganda, for giving them such attention, time, and opportunity to publish this book.

The authors would also like to take this opportunity to express their gratitude to all those people who have provided them with invaluable help in the writing and publication of this book.

Contents

List of Figures

List of Table

About the Authors

Abid Yahya began his career on an engineering path, which is rare among other research executives. He earned his bachelor's degree from the University of Engineering and Technology, Peshawar, Pakistan, in Electrical and Electronics Engineering majoring in telecommunication and his M.Sc. and Ph.D. degrees in wireless and mobile systems from the Universiti Sains Malaysia, Malaysia. Currently, he is working at the Botswana International University of Science and Technology. He has applied this combination of practical and academic experience to a variety of consultancies for major corporations.

Prof. Abid Yahya is a member of the Institute of Electrical and Electronics Engineers (IEEE), USA, and a Professional Engineer registered with the Botswana Engineers Registration Board (ERB). He has more than 120 research publications to his credit in numerous reputable journals, conference articles, and book chapters. He has received several awards and grants from various funding agencies and supervised several Ph.D. and master's candidates. His recent three books are (1) *Mobile WiMAX Systems: Performance Analysis of Fractional Frequency Reuse* published by CRC Press I Taylor & Francis in 2019, (2) *Steganography Techniques for Digital Images* and (3) *LTE-A Cellular Networks: Multi-Hop Relay for Coverage, Capacity, and Performance Enhancement* published by Springer International Publishing in July 2018 and January 2017, respectively, and are being followed in national and international universities. Prof. Yahya was assigned to be an external and internal examiner for postgraduate students. He has been invited several times to be a speaker or visiting lecturer at different multinational companies. He sits on various panels with the government and other industry-related boards of study.

Joseph M. Chuma is a Full Professor of Electronics Systems Engineering at the Botswana International University of Science and Technology (BIUST). He obtained a Ph.D. in Electronics Systems Engineering and an M.Sc. in Telecommunications Engineering and Information Systems from the University of Essex in the UK in 2001 and 1995, respectively; a B.Eng. in Electrical and Electronics Engineering from the University of Nottingham in 1992; and a Master in Business Administration from the University of Botswana in 2010. Before joining BIUST in 2014, he was the Dean of the Faculty of Engineering and Technology at the University of Botswana. He has over 24 years of experience in teaching and research, consultancy, and human resources development in telecommunication, computer, and electrical and electronics engineering, including CISCO computer networking. Professor Joseph Chuma is a member of the Institute of Electrical and Electronics Engineers (MIEEE), USA, The Institution of Engineering and Technology (IET), UK, Botswana Institution of Engineers (BIE), and a Professional Engineer registered with the Botswana Engineers Registration Board (ERB). He is the Vice-Chairman of the Botswana Communications Regulatory Authority (BOCRA). Professor Chuma has three of his patent applications registered with the World Intellectual Property Organization (WIPO) and also holds the UK and USA patents to his credit. He has successfully completed projects that received research grants for over USD 4 million. He has authored or co-authored three books, three book chapters, and many refereed published scholarly/scientific journal articles in the subject of Electronics and Telecommunications Engineering.

Ravi Samikannu is currently working as an Associate Professor in the Department of Electrical, Computer and Telecommunication Engineering at the Botswana International University of Science and Technology, Palapye, Botswana. Ravi Samikannu obtained his Ph.D. in Electrical Engineering from Anna University, Chennai, India. He has 14 years of teaching experience at undergraduate and postgraduate levels. At present 5 Ph.D. and 4 master's students are doing research under his supervision. Three Ph.D. students and 17 master's students completed their research work under his supervision. He has published 60 research papers in international journals. He has presented 40 papers in international and national conferences and has received the best paper award two times for his presentation. Prof. Ravi is presently working on research projects in smart grid, renewable energy system, and power system. He is actively engaged in teaching, research, and academic administration. He is the reviewer of IEEE and other reputed journals and has delivered special guest lectures in many international and national conferences. He is an active member in IDDS and participated in different rural community development projects.

Adamu Murtala Zungeru received his Ph.D. from Nottingham University. He was a Research Fellow at the Massachusetts Institute of Technology (MIT) in the USA, where he also obtained a Postgraduate Teaching Certificate in 2014. He is currently an Associate Professor and the Head of the Department of Electrical, Computer and Telecommunications Engineering at the Botswana International University of Science and Technology (BIUST). Prof. Zungeru is a senior member of the Institute of Electrical and Electronics Engineers (IEEE) and a registered engineer with COREN, ACM, and ERB of Botswana. He is the inventor of termite-hill routing algorithm for wireless sensor networks and has two of his patent applications registered with the World Intellectual Property Organization (WIPO). He has also authored 4 academic books and over 60 international research articles in reputable journals, including *IEEE Systems Journal, IEEE Internet of Things Journal, JNCA*-Elsevier, and others, with over 700 citations and an H-Index of 12. He has also served as an international reviewer for *IEEE Transactions on Industrial Informatics, IEEE Sensors, IEEE Access, IEEE Transactions on Mobile Computing, IEEE Transactions on Sustainable Computing, JNCA*-ELSEVIER, and numerous others. He is presently serving as the Chairman of IEEE Botswana Sub-Section.

Caspar K. Lebekwe was awarded a first class degree of MEng in Electronics and Communications Engineering at the University of Bath in 2008. He holds a Ph.D. in Electrical and Electronics Engineering, sponsored by the General Lighthouse Authorities at the same university. His Ph.D. project was focused on eLoran Service Volume Coverage Prediction. He is currently a lecturer at the Botswana International University of Science and Technology, where he teaches Optical Communications, Antennas and Propagation, Discrete Mathematics, Telemetry, and Remote Control as well as Electromagnetic Field Theory. He currently supervises three Ph.D. students and two master's students. His main research interest centers on Service Volume Coverage Prediction at various frequencies. He is currently working on a 5G network coverage prediction radio tool for Botswana. He is a registered member of the IEEE.

Taolo Tlale holds a bachelor's degree in Telecommunications Engineering having completed at the Botswana International University of Science and Technology (BIUST) in the year 2018. He currently works for Mascom Wireless (Pty) Ltd in Botswana as a Billing Intern. His research areas portrayed by various research projects he has worked on span the Internet of Things (IoT), cloud computing, and wireless sensor networks (WSNs). With expertise in sensor and transducer electronics, near-field communications (NFC), and embedded systems, he focused on agriculture, tourism, weather, and vehicle monitoring systems inclusive of the IoT and cloud computing. He has received research grants for the development of systems in the mentioned areas. Taolo Tlale currently plays the role of a data center engineer at Mascom Wireless (Pty) Ltd for maintenance of storage servers, storage area network (SAN) switches, and backup systems for data centers.

Leatile Marata received a Master of Engineering (MEng) in Telecommunications and a Bachelor of Engineering (BEng) in Telecommunications and Electronics Engineering from the Botswana International University of Science and Technology and Central University of Las Villas, respectively. He is currently working for the Botswana International University of Science and Technology as a Teaching Instructor. His current research interests are in the areas of Bayesian estimation and detection methods for wireless communications, machine-type communications, and molecular communications. He has conducted some research visits to the University of Oulu and University of Loughborough.

Chapter 1
Agriculture: Wireless Sensor Network Theory

Abstract This chapter provides the Internet of Things (IoTs)-based Smart Agriculture System. The system aims at improving agricultural production in Botswana by remotely monitoring farms of all types using the Internet of Things (IoTs). The system uses sensors to monitor different parameters for these living things to ensure that a proper environment is maintained at all times. This chapter gives a detailed explanation of the data loggers which transmit data wirelessly to a central point located on the farm. The central location is referred to as a gateway, and it is where the farm employees can visualize and analyze data. The on-farm network of sensors and gateway is linked to an online server through General Packet Radio Service (GPRS) and Satellite to allow for remote online data acquisition (DAQ). In this chapter, low cost and reliable wireless data acquisition system are implemented in real time at the banana field. The moisture stress, reducing the usage of excessive water, rapid growth of the weeds is achieved with the implementation of sensor-based site-specific irrigation. Internet of Things-based remote control of irrigation can also be achieved in the system. The implemented system can be used to transfer the fertilizer and other chemicals to the field with the help of adding new sensors and valves.

1.1 Introduction

According to Martin and Islam (2012), a wireless sensor network (WSN) is an interconnection of different devices at different locations to monitor their site. The authors stated that the WSNs send information wirelessly through each other using radio signals. The data is sent to those who need it for observation and analysis. Still, according to Martin and Islam, most WSNs are designed to have a sink or base station which collects data from all the monitoring sites. The users may send commands into the network whenever they need data. The monitoring sites are referred to as sensor nodes, and they are usually composed of the sensing devices, radio transceivers, computers at integrated circuit size, and power supplies.

© Springer Nature Switzerland AG 2020   1
A. Yahya, *Emerging Technologies in Agriculture, Livestock, and Climate*,
https://doi.org/10.1007/978-3-030-33487-1_1

1.2   Origins of Sensor Networks

Chong and Kumar (2003) have provided a detailed description of the evolution of WSNs. They mentioned that the first sensor networks were developed during the Cold War era by the military of the United States of America (USA). The sensor networks established by the army at the time were meant to operate at the bottom of seas, and oceans detect the stealth submarines of the then Union of Soviet Socialist Republic (USSR), mostly present-day Russia. This was accomplished by the development of a system referred to as the sound surveillance system (SOSUS) which was a network of hydrophones placed at the bottom of the seas. The SOSUS was followed by the development of distributed sensor networks (DSN) by the Defense Advanced Research Projects Agency (DARPA). The DARPA research was followed by an explosion of the WSNs with a wide range of applications seen today. Different standards have been developed for being implemented on WSNs.

1.3   Sound Surveillance System (SOSUS)

There has always been a need to detect unrecognized submarines. According to Whitman (2005), during World War I, the navy of the United States used sound navigation and ranging (SONAR) to perform the acts of submarine detection. However, still according to Whitman (2005), the SONAR used in World War I had the disadvantage of short-range detection; hence there was a need for a system that would detect the submarines at longer ranges. In 1937, a scientist named Maurice Ewing from Lehigh University discovered while doing experiments with sound in the Atlantic Ocean that sound waves can propagate over long distances underwater with less attenuation. The scientist hypothesized that these sound waves could be detected thousands of kilometers away from the source using the necessary hydrophones. This discovery by Maurice Ewing was supported by the development of an instrument called the *bathythermograph* at the Massachusetts Institute of Technology (MIT) and the Woods Hole Oceanographic Institution. The bathythermograph is an instrument used to provide a continuous measurement of ocean temperature as a function of depth. Thanks to the bathythermograph, the temperature profile of an ocean could be determined. This instrument also provided an insight into the velocity of sound waves propagating horizontally at different depths of the sea. Later a discovery was made by these researchers that there exist a layer at the bottom of an ocean that can trap sound waves and allow them to propagate horizontally without hitting the ocean floor nor its surface. This was due to a realization that at the surface of the sea where the water is warmer, the sound will propagate at a higher velocity. As the depth increases, the sea temperature decreases, and so does the sound velocity, it decreases. Going further down the sea, the effect of pressure due to the water causes the velocity of sound to increase again. This formed the sound velocity profile (SVP). Another phenomenon is that a sound wave coming from a low-velocity region to a high-velocity region gets reflected into the lower velocity region. Using the SVP and the reflection phenomenon, it was discovered that there exists a layer in which sound velocity is minimum and will allow the waves to propagate horizontally. This layer that traps the sound waves were given a name the *Deep Sound Channel*. Sound waves coming from any other depths will be stuck in the deep sound channel once they reach it.

Further experiments by Ewing on this discovery led to the development of a system known as Sound Fixing and Ranging (SOFAR) in 1944. This was a rescue system meant for pilots having fallen on to seas. The principle was to have the pilots drop explosives which would detonate at the deep sound channel, the sound of which would be detected by hydrophones. The position of the source of the explosion would be determined by triangulation and subsequently that of the pilot. This would be helpful during the ongoing World War II. At the end of the war, research by the US navy on underwater acoustics continued, with improvements made on the SOFAR in conjunction with the Air Force.

The end of World War II saw the uprising of a competition between the USA and the USSR in technology, a race referred to as the Cold War. The USA in the early 1950s concluded that the greatest threat to its security was the USSR submarines. Therefore the focus of research on underwater acoustics was redirected to detecting these submarines. Project Hartwell based at MIT was established to reach the desire of long-range submarine detection. The project proposed that the already existing SOFAR can be used for long-range submarine detection. Project Hartwell also realized that sound at frequencies below 500 Hz could reach the deep sound channel from any depth of the sea. In 1950, the US Navy contracted the American Telephone and Telegraph Company (AT&T), Western Electric and Bell Telephone Laboratories to start working on the desired long-range submarine surveillance system. The prototypes of this surveillance system were made up of hydrophones in a single line and placed at the bottom of the ocean. This surveillance system would then be named the sound surveillance system (SOSUS). The signals picked up by the hydrophones were transmitted by cables to signals processing facilities referred to as Naval Facilities (NAVFACs). In the NAVFACs, trained personnel would observe the received signals to distinguish the difference between the sound due to a submarine and that of the ocean background, as explained by Whitman (2005). The SOSUS continued to operate from the early 1950s until its effectiveness declined during the 1970s due to the improvements made on the USSR submarines.

1.4 Distributed Sensor Network

The DARPA is said to have ventured into sensor networks in the late 1970s, according to Nack (2008–2009). The DARPA found an interest in tracking and surveilling distributed systems and processes. For this reason, they came up with the distributed sensor network, which was made up of sensors, databases, and processors. The uses of the DARPA distributed sensor network included surveillance of aircraft. That of the universities followed research by DARPA as they developed WSNs for applications in the areas of monitoring air quality, forest fires, natural disasters, weather, and structures (e.g., bridges). Martin and Islam further described the applications of WSNs by grouping them as military, area monitoring, transportation, health, environmental monitoring, industrial monitoring, and agricultural monitoring. In this project, an application of WSNs on animal health is demonstrated.

The impossible has become achievable in Africa due to the fastest growth in the mobile telecommunication industry. Africa, the second biggest mobile market, has sparked several digital innovations, with diverse applications in the sector of education, health, ICT, agriculture, governance, finance, energy, and tourism.

The rapid growth of tech hubs in Africa gives birth to digital entrepreneurship ecosystem by networking entrepreneurs, designers, developers, and prospective investors.

1.5 Rural Households in Sub-Saharan Africa

In sub-Saharan Africa, rural households do not bound labor provision to agriculture; however, also function and labor in non-farm enterprises (Reardon et al. 2006). With time the household earnings and employment have improved instead of cut due to the contribution of these enterprises (Haggblade et al. 2010; Lanjouw and Lanjouw 2001; Start 2001). This involvement will make the Africa labor market attractive for the approximately 170 million new job hunters between 2010 and 2020 (Fox and Pimhidzai 2013).

Nagler and Naudé (2017) have conducted a study in six sub-Saharan African countries in terms of labor efficiency, existence, and withdrawal, using the World Bank's "Living Standards Measurement Study-Integrated Surveys on Agriculture" (LSMS-ISA). It shows from the study that those who are far from the central market and businesses have lower levels of labor output related to urban and male-owned enterprises. Furthermore, a connection and association have been suggested by authors between a household's incentive to run an initiative and its successive progress (Nagler and Naudé 2017).

1.6 Farming Systems Research (FSR) in Africa

Farming systems research (FSR) became an established method in the late 1970s particularly to appreciate the objections of interpreting a green innovation of agricultural modernization into the heterogeneous developments atmospheres of Africa and Latin America (Collinson 2000; Norman 1995).

Whitfield et al. (2015) highlighted agricultural involvements problems in eastern and southern Africa. Authors studied modifications in the method and conceptions of FSR and indicate the worth of farming systems perceptions that go further than these accumulations, and suggest a solution to capture the multilevel system changing aspects that bond on-farm policy-making to extensive social, political, and ecological ups and downs.

1.7   Conservation Agriculture

Conservation agriculture (CA) goals to utilize agricultural means by putting into practice of least soil disruption, everlasting soil protection, and crop divergence simultaneously (Thiombiano and Meshack 2009).

The eminence of CA is endorsed by the extraordinary prospective for improved agronomic and ecological effects (Ndah et al. 2014; Thierfelder et al. 2015, 2016; Mupangwa et al. 2016).

In Africa, during the era between 2008/2009 and 2013, the area under CA is claimed to have increased by 57% to more than 1.2 million ha (Kassam et al. 2015).

The conservation agriculture (CA) is promoted intensively to African smallholder farmers. However, the technologies and their implementation are both defined below par, directing to a large deviation in approximations and rationality concerns (Brown et al. 2018). Brown et al. have applied Conservation Agriculture Appraisal Framework (CAAF) and Process of Agricultural Utilisation Framework (PAUF) to household survey data from 1601 villages and 6559 households over five eastern and southern African countries. In the study, authors have found a common overvaluation of CA approval and CA mechanisms. The proposed structures have displayed strong prospective to deliver an improved complexity of sense in the measurement of approval and non-approval.

1.8   Modernization of Agriculture

Yang and Zhu (2013) developed a two-region model that highlighted the essential role played by agricultural modernization in the evolution from stagnation to development. Authors pointed out that slow productivity growth in the industry is linked with traditional agriculture, where the extra labor force is needed to produce food, hence grand a limit on rising per capita income. Authors further added that agricultural modernization flares up the evolution to new growth.

For instance, when structural revolution speeds up alongside with the improvement of agriculture, the profitability to human capital is probably to increase. Therefore, families will concentrate on human capital rather than having more children.

1.9 Application of Internet of Thing (IoT) and Smart Farm

Information and Communication Technologies (ICTs) play an essential role in refining competence, value, and throughput; instead, ICTs are not used well in agriculture (Milovanovic 2014). Slight modifications in productivity can have a foremost influence on turnover (Bewley et al. 2015). Most important to effectiveness is the combination of mixed sources of facts in an attempt to permit economically possible and environment-friendly decision-making.

Technologies, coupled with sensors, facilitate farmers to monitor their farms in real time. This suggests an interesting option of mounting farm-targeted models that one can practice to design their actions in reaction to shifting conditions (Michael and Gregory 2017).

Technologies supporting modern farms provide an opening for assisting the building and application of farm-specific models (Michael and Gregory 2017).

The Smart Farm (Wark et al. 2007) is established on extensive heterogeneous sensing. Heterogeneous sensors can extend a variety of techniques, and be incorporated in use from manufacturers coupled with several algorithms, hardware, and protocols. Although heterogeneity has an enormous impact, dealing with it is thought-provoking. A mutual methodology is that of middleware, which suggests appropriate points of concepts such that the heterogeneity in its different scopes can be moderated and efficiently accomplished (Bhuyan et al. 2014; Lee et al. 2012; Wang et al. 2008). SIXTH (O'Hare et al. 2012), a distributed middleware infrastructure, enables and supports this type of sensor network. An open-source sensor, a Global Sensor Network, is being proven as a facilitator of the modern farm (Gaire et al. 2013).

Internet of Thing has contributed significantly in both diagnostics and control in the field of environment and agriculture industry. Moreover, IoT can offer data to the end user concerning the product starting point and its effects (Talavera et al. 2017). To improve the productivity of agriculture, deploying a wireless sensor network (WSN) in the field has brought better quality and competence to the farmers (Capello et al. 2016; Fang et al. 2014; Hashim et al. 2015; Kodali et al. 2014). WSNs assess soil moisture, weather situations, temperature, humidity, vibrations, and biomass of plants or animals (Pang et al. 2015). Besides, WSN can be used as real-time monitoring of crop growth and productivity. Moreover, WSN defines the ideal time to harvest, discover diseases, and other issues related to agriculture and farmer (Ndzi et al. 2014).

Fang et al. (2014) presented a novel integrated information system (IIS) coupled with modern tools to study the ecological effects of climate change in the region of Xinjiang. An evident growing inclination of the air temperature in Xinjiang over the last 50 years has been observed from the case study.

Li et al. (2013) proposed an IoT-based distributed information service system of agriculture. Authors designed information-discovery system to secure, regularize, locate, handle, and get business data query from agriculture production. In this study, a coding scheme was used to identify the agricultural product. Moreover, distributed IoTs servers were used to trace the agriculture products to confirm the authenticity and value of agriculture products.

Capello et al. (2016) applied the Industrial Internet of Things (IIoT) technology to estimate the effect of real-time monitoring tools in the agriculture sector in terms of quality and effectiveness.

Tzounis et al. (2017) discussed the importance and applications of IoT, Cloud Computing, Fog Computing, and WSN in the agriculture sector.

1.10   Big Data in Agriculture

Agro-environmental investigation use cases generally concern dynamic systems with complicated connections between living organisms or consumable products and their environment. The complexity of describing, analyzing, and accepting such structures and the extent and heterogeneity of the information used can be assumed (Lokers et al. 2016). The complication of managing big data is often termed as the "3 V's" of dig data, such as volume, variety, and velocity (Laney 2001).

Lokers et al. (2016) proposed a theoretical framework to structure and investigate data-intensive cases associated with big data usage. Authors recommended that big data research, and mainly the struggles to be carried in the agro-environmental domain, emphasis should be on diversity and integrity challenges.

Chi et al. (2016) analyzed and described the most challenging concerns in managing, processing, and effective use of big data and the prospects that big data bring in the context of remote sensing applications.

Authors studied two cases of remote sensing data; in the first scenario, marine oil spills were identified automatically. However, in the second case, high-performance computing (HPC) was used performed using to extract information from an extensive database of remote sensing images.

Gutiérrez et al. (2008) used the competence of logistic regression (LR) coupled with product-unit neural network (PUNN) model in the different fields of study to analyze multispectral imagery for forecasting R. segetum existence probability and mapping R. segetum patches. It has been proved from the experiments that the better accuracey were obtained as compared to the traditional RS classification methods.

1.11   Agricultural Monitoring and Irrigation Control System

Kittas et al. (2003) presented a climate model to forecast the air temperature distribution inside a longitudinal greenhouse which integrated the result of ventilation rate, roof shading, and crop transpiration. Authors calibrated the climate model by placing the temperature in the middle and at the end of a significant length of the greenhouse. It has been proved from the results that the cooling scheme was capable of retaining the greenhouse air temperature at slightly low levels.

Kacira et al. (2005) recommended that response-based sensing for control approaches in ecological greenhouse production is very important. Authors suggested that noncontact real-time data monitoring is required to check the actual status of the plants without any physical interference.

Verdouw et al. (2013) presented a hypothetical framework for the investigation of virtual supply chains from the IoT viewpoint, which goes further than the leading organizational perception of virtual supply chains. Authors added a methodical application of the perception of virtual things to the domain of supply chain management (SCM). Lastly, the proposed conceptual framework has been applied in the Dutch floriculture for analysis. It has been shown from the analysis and study that future issues are not only involved with technology problems, nevertheless specifically similarly with functional and organizational matters.

Nishina (2015) designed a chlorophyll fluorescence imaging robot to observe the fitness situations of tomato crops in the semi-commercial greenhouse by determining the induction curves. It has been proved from the design that there was an obvious heterogeneous distribution of photosynthetic functions diagonally the 20 m X 11 m farming area. Authors recommended that to regulate the tomato fruit color with storage, tomato fruit color assessment practice can be added.

Aiello et al. (2018) proposed a methodological approach to support the farmers in executing more supportable and effective production developments, consistently with the principles of the biobased economy.

In this study, the proposed methodology encouraged a knowledge-based bioeconomy as a technological help to increase the efficiency, competence, and sustainability of biological processes. Authors recommended that decision support systems (DSS) can be an actual solution to upkeep small-scale farmers in their decision-making processes, and thus increasing productivity.

Gutiérrez et al. (2014) designed and developed an automated irrigation system for crops employing WSN and GPRS module. The solar power designed system proved that the usage of water could be reduced for a certain quantity of fresh biomass production.

Fourati et al. (2014) developed a web-based weather station for irrigation scheduling in olive fields communicating with WSN. To calculate the water crop requirement precisely, solar radiation, humidity, rain, and temperature sensors were deployed.

Luan et al. (2015) proposed an integrated service for agricultural drought monitoring and anticipating, and irrigation control on IoTs-based system.

Karim et al. (2017) proposed IoT-based alert system for the control of water stress of plants to be able to approximate the amounts of water needed.

Mohanraj et al. (2016) presented an e-Agriculture Application based on the structure comprising of KM-Knowledge base and monitoring sections. Authors proposed an evapotranspiration system to compute the water necessity of a plant per day with the developed algorithm.

A WSN-based best possible use of watering crops control system is developed with data management through mobile phone and a web-based application and applied in Makham Tia District, Surat Thani Province, Thailand. It is found that the moisture amount of the soil was sustained properly for vegetable growth, cost not only effective but also growing agricultural yields (Muangprathub et al. 2019).

1.12 Farm Management Systems

Zhao et al. (2010) developed a remote monitoring system for greenhouse-site based on IoT and designed an agriculture information management system.

Husemann and Novković (2014) emphasized on tangible friendly user Farm Management Information Systems (FMIS) that the farmer has to be allowed to assign the unusual means of the farm. Authors analyzed German ranch from the state North-Rhine Westphalia as a case study.

Authors developed an easily adaptable, user-friendly, and precision in describing different production process FMIS that ensembles the requirements of the case-study ranch. Authors have concluded that basic FMISs offer a satisfactory broad structure; nevertheless, a lot of modifications are required in case of an application on the real ranch to describe all production developments precisely.

Nizam and Sazili (2015) proposed a web-based FMIS for smallholders in Malaysia, employing rapid application development (RAD) prototyping methodology. The final system has been developed, followed with Malaysian Good Agriculture Practices (MyGAP) compliance and implemented on the official MyAgris website [http://www.myagris.com].

Canfarm is the first cooperative FMIS used by Canadian farmers for recordkeeping and planning services (Thompson 1976). Kok and Gauthier (1986) then presented a user-friendly farm information management system installed on a microcomputer. The architecture of the designed system consists of the processing of permanent data, annual data link, daily farm operations, and inventory data. This kind of prototype is flexible and still in use in many commercial applications.

Fountas et al. (2015) reviewed up to date FMIS from both an academic and commercial viewpoint. Authors highlighted on open-field crop production and analyzed FMIS results for precision agriculture as the most prominent data demanding application area.

Paraforos et al. (2016) developed an FMIS precision agriculture based on Future Internet Public-Private Partnership Program (FI-PPP). Authors emphasized the on-farm financial analysis, the furthermore developed App capable of assessing the fixed values of farmer imports.

Questionnaire Survey

Botswana is a country that is made up majorly of the agricultural sector, among other yet very few areas. Although this is the case, this sector is one of the stagnant and fast deteriorating regions due to factors of climate change and diseases of livestock and crops, to pick a few. The rapid decrease in agricultural activities is bound to cause a significant blow to the economy of Botswana; this is mainly because the major export in Botswana is beef.

So with this in mind, researchers have aimed to work together with the community and government structures to come up with technologies that will help cultivate the quality of agricultural activities in Botswana. To collect information on the matter questionnaires were distributed from which the results are analyzed.

Fig. 1.1 Pie chart

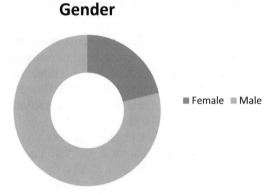

Gender

■ Female ■ Male

Fig. 1.2 Age group

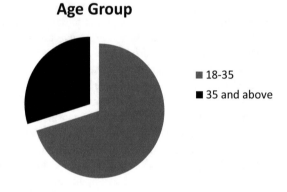

Age Group

■ 18-35
■ 35 and above

Fig. 1.3 Nationality

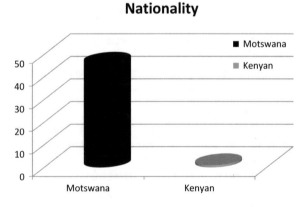

Nationality

■ Motswana
■ Kenyan

From the pie chart, as shown in Fig. 1.1, it shows that most of our questionnaire participants were men. The actual numbers are 37 males and 10 females. It suggests that agriculture is a male-dominated sector, as most of the willing respondents were males.

The dominant age group is 18–35 with 33 people and the less group of participants being of ages 35 and above with only 14 people, as shown in Fig. 1.2. A lot of

Farming District

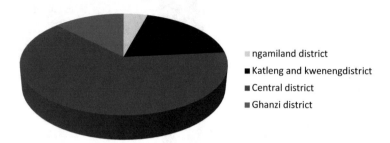

■ ngamiland district
■ Katleng and kwenengdistrict
■ Central district
■ Ghanzi district

Fig. 1.4 Farming district

Fig. 1.5 Occupation

Occupation

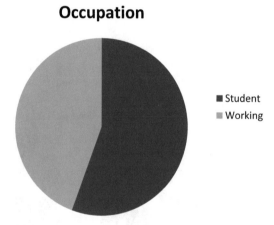

■ Student
■ Working

the respondents were the youth, and this shows that unlike before more young people are involved in agriculture.

Figure 1.3 shows that only one from 47 of the respondents was not a Motswana.

Figure 1.4 shows most of the respondent's farm around the central district.

The majority of respondents were from students with 25, and those working were 22, which is not a significant margin, as shown in Fig. 1.5.

1. *What type of farming are you involved?*

    Most of the respondents were subsistence farmers (37), a few in commercial farming (8), and only two were practiced both as presented in Fig. 1.6.
2. *Which sector of agriculture are you pursuing?*

    Most of the respondents are into animal production, and a few in crop production and a smaller number of the respondents practice both. Nine respondents were in crop production, 30 in animal production, and 8 in both sectors, as shown in Fig. 1.7.
3. *What is the predominant problem farmer's face in Botswana?*

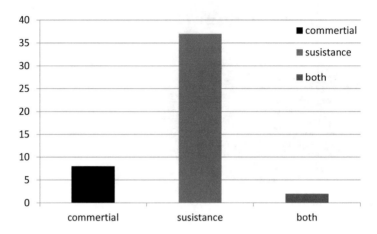

Fig. 1.6 Types of farming

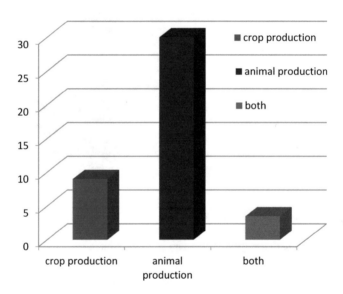

Fig. 1.7 Agriculture sector

A large number of participants felt that the leading problems are climate change and lack of knowledge and poor farming practices as depicted in Fig. 1.8.

4. *Do you know of any technologies developments aimed at helping farmers in Botswana?*

From all the respondents, a lot of them do not know of any technological developments aimed at helping farmers in Botswana that is 30 of the respondents as shown in Fig. 1.9. Those who knew (17 respondents) mentioned free

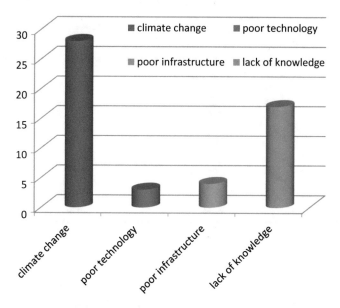

Fig. 1.8 Predominant problem farmer's face in Botswana

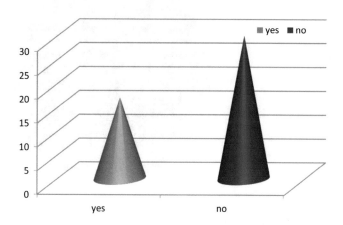

Fig. 1.9 Technologies developments aimed at helping farmers in Botswana

vaccination, irrigation, blue economy, ISPAAD, LIMID, NUMPAD, ear tag system (bolus), CEDA, etc.

5. *Is the necessary information about your livestock or crops readily available to you?*

   Twenty-three respondents answered no, and 24 answered yes, which gives almost a 50/50 response, as shown in Fig. 1.10.

6. *Would you be interested in using an agricultural-based application with information on livestock, crops, and updated information from organizations in agriculture?*

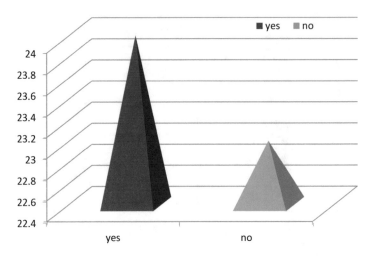

Fig. 1.10 Information about livestock or crops readily available

Fig. 1.11 Agricultural-
based application with
information

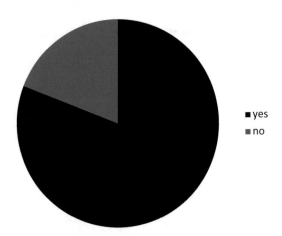

Figure 1.11 shows that 80.1% of the respondents are interested in acquiring
an agricultural application for easy access to information. Specification requirements included multi-language, farming equipment, microchip devices, tracing
livestock, low cost, disease detector, easy to use, information on climate change,
reliability, and offline functionality.

7. *Have you ever tried any methods to improve your farming? Either self-imposed
   or proposed by organizations or government?*

Figure 1.12 shows that 28 of the respondents have never tried any practices
to improve their farming. The 16 who had have mentioned the use of high breed
seeds, NUMPAD, zero grassing, feedlot, infrastructure improvement, row
planting, irrigation, crop spacing, and most argued that the methods were
beneficial.

Fig. 1.12 Methods to
improve farming

■ yes

■ no

8. *When was the last foot and mouth outbreak in your area and how long did it last?*

   The responses ranged from an outbreak from between 2007 and 2018 lasting between 3 months and 2 years. Some of the respondents are from free zone areas.

9. *What fraction of your livestock have you lost to foot and mouth disease?*

   Most of the respondents who experienced the outbreak lost from half of their livestock downwards.

10. *Where have any measures taken to control or prevent foot and mouth in your area? If yes, please note them.*

   Free vaccination, veterinary gates, and livestock movement restrictions were the few that were mentioned by the few respondents who got assistance.

11. *Do you know of any technologies or systems addressing climate change as a significant effect in agriculture? If yes, please state them?*

NPK fertilizers, seed-co-seeds that withstand harsh weather, and greenhouse were the few that were mentioned by less than 5% of the respondents. The rest did not know of any.

The information from the questionnaires suggests that the central focus should be in educating farmers to improve their farming practices, developing affordable borehole drilling, government assistance for all subsistence farmers, and new technology updates being readily available to farmers.

Smart Farm Monitoring System Employing the Internet of Things (IOT)

The Internet of Things (IoT) has already been used to set the pace in agriculture by Savale et al. (2015) as they designed a wireless sensor network (WSN) for precision agriculture. According to the authors, the system applies to both farms and greenhouses. The architecture consists of a central system, control cabinet, and sensor nodes. The primary system consists of a communication server, database, and web server. The sensor nodes pick up data which are communicated to the central system via the control cabinet. The control cabinet provides access to the sensor nodes. The control cabinet is also connected to a gateway that provides the farmer with data. An agronomist is alerted by the central system in case an abnormal event occurs. The system does not provide real-time monitoring as the agronomist is only alerted upon an abnormal event. Keshtgari and Deljoo (2012) designed a WSN solution for precision agriculture based on ZigBee technology. The system employs sensor nodes deployed in the farm and uses ZigBee radios to transmit data to a gateway that is connected to a remote server. The WSN uses grid topology having the gateway as the sink node at the center of the grid. The sink node transmits data to a BS every 15 min. The time interval of 15 min for transmission of data to the BS means there is no real-time updating of information regarding the farm conditions.

A WSN for greenhouse parameter control was designed and tested by Chaudhary et al. (2011). They tested the system in a 35 m by the 200 m greenhouse with four types of sensors to monitor climates outside and inside the greenhouse, soil conditions being moisture, temperature, pH, and electric conductivity and the last sensor type monitoring greenhouse environment. The greenhouse has actuators for the regulation of humidity, air temperature, irrigation, and daylight. A precision farming (PF) solution based on WSN in Egypt was proposed by Abd El-kader and El-Basioni (2013). The prototype was designed to monitor soil parameters in a potato field. In testing the prototype, the authors evenly distributed the sensor nodes throughout a potato field. The sensor nodes consist of sensors for humidity, soil moisture, temperature, pH, soil, and light intensity.

The node is completed by attaching the sensors to the MDA300 DAQ board and a MICAz radio platform. The WSN uses Periodic Threshold-sensitive Energy-Efficient sensor Network (APTEEN) routing protocol. The sensor node clusters are connected to the Internet or cellular network to allow for remote monitoring. Since the WSN was designed only for monitoring a potato field, it may not apply to other crops. Kassim et al. (2014) designed a novel Intelligent Greenhouse Management System (IGMS) based on a WSN. The IGMS has temperature, moisture, and humidity sensors. The system allows for remote monitoring as the data is transmitted to a server. The system can control actuators that perform certain tasks in the greenhouse-like automated irrigation.

The proposed WSN provides monitoring of only three parameters, which is not satisfactory for large-scale production. Roham et al. (2015) developed a greenhouse monitoring system using a WSN. They monitor the conditions of carbon dioxide, temperature, and humidity. ZigBee sensing nodes are deployed in the greenhouse to take the measurements. The system relies on ZigBee routers and gateway for gathering data. The gateway sends data to a BeagleBone board which is connected to the Internet to transmit data to a database. The system monitors only three parameters which are not enough to ensure a highly productive farm. A WSN was designed and implemented by Shen et al. (2016) for livestock house environmental perception monitoring. The WSN architecture comprises sensor nodes deployed in livestock houses, routing nodes for collecting data from the sensor nodes, and the central node that relays data to a server. The sensor nodes have sensors for temperature, illumination, ammonia gas, sulfuretted hydrogen, humidity, and carbon dioxide. The sensor nodes and routing nodes are based on the ZigBee standard. The WSN uses GPRS for data communications with the server. Because of the use of the only

GPRS for data communications, the WSN system will not be able to upload data to the server, hence limiting remote monitoring.

1.13   Farm Monitoring System

Real-time integrated farm monitoring system using the Internet of Things (IoT) is a project to improve agricultural production by remotely monitoring farms of all types using the IoT where the IoT refers to a set of protocols being developed to allow objects to have an Internet connection for remote monitoring and control through the Internet. The system employs a network of sensors and wireless communication standards to monitor different conditions in both crops and animal environments. A typical integrated farm encompasses an animal field and a crop field. Examples of animals kept in farms in Botswana are cattle, pigs, small stock and poultry or chickens while a wide variety of vegetables, fruits, and cereals are produced. For a high-quality production of these living things, a suitable environment must be provided at all times. The proposed system uses sensors to monitor different parameters for these living things to ensure that a proper environment is maintained at all times. In case a particular parameter reaches a threshold level, the system alerts the farm employees so that they can make the necessary adjustments. A typical integrated farm needs to have the parameters of temperature, humidity, light, and gas concentrations of carbon dioxide, ammonia, and hydrogen sulfide tracked by sensors for animals. These conditions may also be used to detect the presence of parasites and pests if the levels of conditions that favor their growth are known. The conditions being monitored for crops may be soil moisture, pH, temperature and nutrients or fertilizer concentration. The system also keeps track of atmospheric conditions such as temperature, rainfall, pressure, humidity, wind direction, and speed. Depending on the levels of these conditions, the necessary actions may be taken to keep the animals and crops in good health. The measures to be taken may be watering or irrigation, application of fertilizers, application of insecticides and pesticides, provision of feeds to animals, and regulation of temperatures in animal housing, e.g., poultry house. In the system, the sensors are connected to data loggers which transmit data wirelessly to a central point located on the farm. The central point is referred to as a gateway, and it is where the farm employees can visualize and analyze data. The on-farm network of sensors and gateway is linked to an online server through General Packet Radio Service (GPRS) and Satellite to allow for remote online data acquisition (DAQ).

In Fig. 1.13, the sensors measure different parameters, which are then transmitted by data loggers wirelessly to the gateway. The gateway serves as the center of the system. The farm operator fetches data from the gateway either by connecting to it using an Ethernet connector if they happen to be near it or by wireless means if they are at a distance. Figure 1.24 provides some more details on how the system operates.

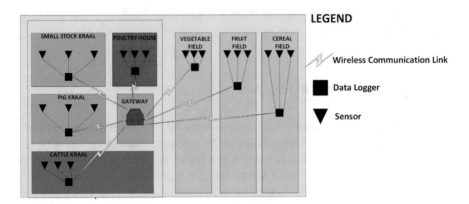

Fig. 1.13 An example of an integrated farm being monitored by the proposed system

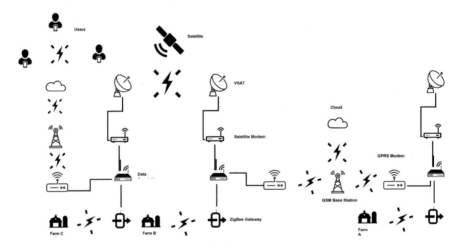

Fig. 1.14 The communication network of the system monitoring different farms and linked to the Internet for remote monitoring

In Fig. 1.14, each farm is equipped with satellite communication equipment and a General Packet Radio Service (GPRS) modem. The sensors deployed in the farms use ZigBee IEEE 802.15.4 standard for data transmission to the gateways. The system is proposed for relaying data from other farms which may be far away from GSM infrastructure. Both GPRS and communication satellites are connected to the Cloud network, which has servers for storing data about the farms. A user can access this information from anywhere in the world. The system would also allow for autonomous performance of activities such as automatic irrigation or watering and temperature regulation if in case there are actuators for that in the farm.

Fig. 1.15 Collecting Data using the Questionnaire and Interview

Fig. 1.16 Farming practices in the north-west district of Botswana

Types of Farming in North-West District, Botswana

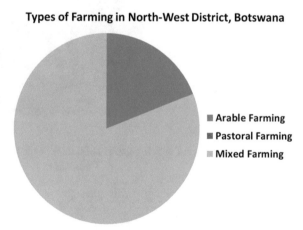

- Arable Farming
- Pastoral Farming
- Mixed Farming

1.14 Data Collection and Field Trip

The field trip was taken to the northern parts of Botswana to collect data in farms of those regions. In the data collection process, the literate farmers were given the questionnaires to fill while the researchers interviewed the illiterate farmers and filled the polls for them as shown in Figs. 1.15, 1.16, 1.17, 1.18, 1.19, and 1.20. The following is a list of Botswana villages which encompass the farms that were visited (Table 1.1).

Fig. 1.17 Crops produced in the north-west district of Botswana

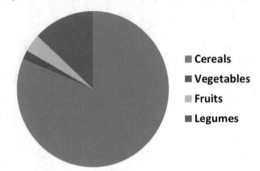

Crops Produced in North-West District, Botswana

- Cereals
- Vegetables
- Fruits
- Legumes

Fig. 1.18 Animals produced in the north-west district of Botswana

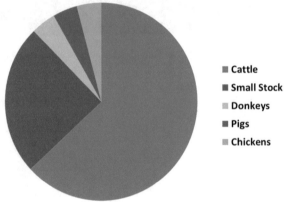

Animals Reared in North-West District, Botswana

- Cattle
- Small Stock
- Donkeys
- Pigs
- Chickens

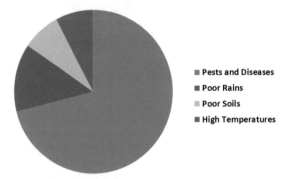

Factors Influencing Crop Production in the North-West District, Botswana

- Pests and Diseases
- Poor Rains
- Poor Soils
- High Temperatures

Fig. 1.19 Factors affecting arable farming in the north-west district of Botswana

Fig. 1.20 Factors affecting
the rearing of animals in
the north-west district of
Botswana

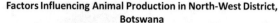

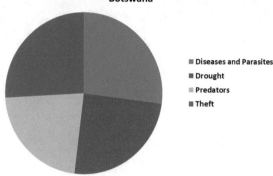

Table 1.1 List of Botswana
villages

No.	List of Botswana villages
1	Tonota
2	Mathangwane
3	Sepako
4	Maposa
5	Nata
6	Zoroga
7	Maun
8	Shorobe
9	Rakops
10	Motopi

From the questionnaires, data were organized according to the responses given by the farmers. This was according to the farming types, crops produced, reared animals, as well as the challenges faced by farmers in the north-west region.

1.15   Technology Readiness

The research was undertaken, and the findings were used to come up with the project thus here. The other principles observed were those in telecommunication system design. In this case, a WSN was being designed and integrated with Cloud computing. The necessary principles for doing this work were observed and followed.

The concept used to solve the identified problem employs IoT and Cloud Computing. Therefore as these are the areas of technology under telecommunication systems, a technology concept got formulated. This included the use of sensors for picking up information, the processing of information, and the transmission of

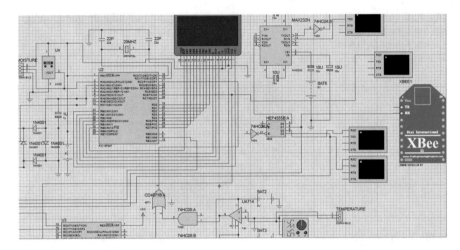

Fig. 1.21 Circuit for the system

Fig. 1.22 The components of the designed network

the information to the necessary places. This is the concept used to address the problem.

The sensor nodes were created and tested in the sense of seeing if they can receive and transmit data as was the wish. These experiments proved to work when the nodes were connected by wired means. The experimental tests were done in the laboratories using the necessary instruments and actual components. The arranged circuits worked very well; Fig. 1.21 shows the circuit diagram of the system. Figures 1.22 and 1.23 show the designed sensor nodes and gateway together with their power supply boards and solar panels in a laboratory after fabrication and integration.

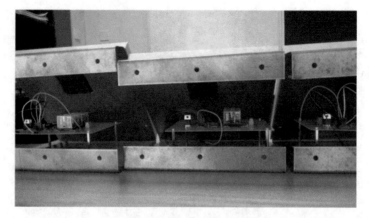

Fig. 1.23 The components under a casing and solar panel mount

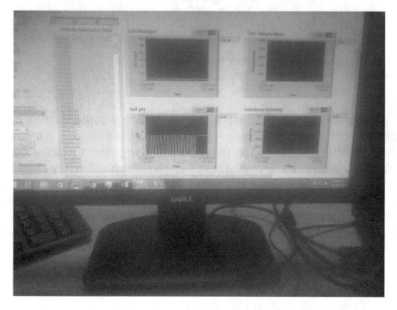

Fig. 1.24 LabVIEW used in displaying data coming from the nodes during an experiment

The sensor nodes were interfaced to a computer to visualize the data coming from the nodes. Figures 1.24 and 1.25 showed an instant when the nodes were being tested in this regard.

Intelligent Wireless Automatic Irrigation System Using GSM
The study of the use of water to the land or soil is called a water irrigation system. The study is utilized to aid the development of farming harvests, support of scenes

Fig. 1.25 Circuit and LabVIEW together for data acquisition

and revegetation of aggravated soils in dry territories and during times of lacking precipitation. Moreover, the irrigation system likewise has a couple of different uses in yield generation, which incorporate ensuring plants against ice, smothering weed developing in grain fields and aiding in avoiding soil combination (Cao et al. 2007). Interestingly, an agribusiness that depends on direct precipitation is alluded to as downpour sustained or dry land cultivating. Irrigation frameworks are likewise utilized for residue concealment transfer of sewage and in mining. Irrigation is regularly contemplated together with waste, which is the regular or counterfeit evacuation of surface and subsurface water from a given region.

1.16 Types of Irrigation Systems

There are mainly five kinds of irrigation facilities. The first one is surface irrigation. The second one is localized irrigation, which has a drip irrigation system. The third one is sprinkler irrigation, which has two types of techniques: one is the center pivot and the other is a lateral move irrigation system. The fourth one is the subirrigation, and the final one is manual irrigation. In the manual irrigation water is supplied with the help of watering cans.

1.17 Drip Irrigation

Drip irrigation is also called a dribble water system. The water is delivered to the roots of plants drop by drop. This method can reduce the usage of water for irrigation. The water can be managed properly because the evaporation is reduced in the drip irrigation system (Shock et al. 1999).

The drip irrigation system can use high technology computerized system to low technology and labor-intensive technology. Low water pressures are needed when compared with other types of systems. The system is using low energy, and it can be designed for efficient usage of water delivery to individual plants. It is challenging to regulate the pressure on steep slopes. Highly calibrated emitters are located along the tube lines that extend the set of valves.

Generally used to grow vegetables and fruits, this framework comprises perforated channels that are put by columns of harvests or covered along their root lines and discharge water legitimately onto the yields that need it. Accordingly, dissipation is decreased, and 25% of water system water is saved in contrast with the flood water system. Trickle water system likewise enables the cultivator to alter a water system program most gainful to each yield. Figure 1.26 shows the system model for drip irrigation.

1.18 Drip Irrigation Tubing

Trickle water system tubing is one of the most significant parts of a dribble water system framework. It is in charge of conveying water to the sprinklers, producers, and sirs. Without it, the water stream to plants is cut off. There are a few kinds of

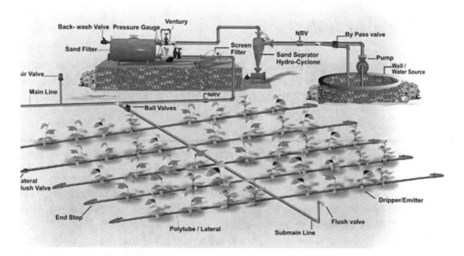

Fig. 1.26 System model for drip irrigation

dribble water system tubing each with its particular function. The most well-known classes are ½, ¼, Soaker, Emitter tubing. Introducing the best possible dribble water system tubing for your trickle water system framework is critical.

1.18.1 1/2 Inch Delivery Tubing

Main water supply is distributed through the 1/2 inch. It keeps running from the tap to the 1/4 inch appropriation tubing, soaker tubing, and producer tubing. It tends to be utilized either above or underneath the soil. Different types of lengths as 50, 100, and 500 feet are available.

1.18.2 1/4 Inch Delivery Tubing

1/4 inch conveyance tubing is utilized to associate the 1/2 inch tubing to the sprinklers, producers, and sirs.

1.18.3 Soaker Tubing

Soaker tubing is like a soaker hose. This system is associated with the trickle water system tubing to the 1/2 inch supply line utilizing the 1/4 inch pointed connector.

1.18.4 Emitter Tubing

Emitter system of tubing is the trickle water system tubing utilized for spot watering, structured explicitly to put an emitter every 12 inches. Similar to the soaker tubing, you have to associate it to the supply line with the 1/4 inch spiked connector.

1.19 Applicable Fields of the Drip Irrigation System

Ordinarily, a drip irrigation framework is utilized to water important plants (vegetable, natural product, and so forth.) that develop in light soil in locations that normally need water or have low-quality water.

It is especially useful when watering blooming plants because, conversely, watering using a sprinkler framework may make plants lose their blossoms since they are so touchy to outside elements. Generally speaking, affordable use of water is exceedingly exhorted in rainless locations, which makes dribble water system frameworks exceedingly attractive when creating this sort of horticultural harvest.

Trickle irrigation frameworks can be connected effectively to open and shut hot house fields, organic-products farms, and developing items that have high monetary worth. Trickle irrigation frameworks are adaptable and can be connected to such different situations as parks, interstates, exchange offices, and houses. This sort of structure is utilized solely in organic products farms, and of late, it has turned into the favored technique for a water system in vineyards to guarantee that there is no an enormous issue or loss of proficiency on the grapes. The pressure danger of a potato plant is essentially diminished when a dribble water system framework is utilized, contrasted with different frameworks. A couple of assets demonstrate that trickle water system frameworks can even be used to develop sugar beets and wheat; notwithstanding, these plants are not exhorted because of restrictive expenses in utilizing the dribble water system framework alongside the way that the yields would need to be unusually thick.

1.20   Advantages of the Drip Irrigation Method

- Facilitates cultivating by using unassuming water applications without risking plant pressure.
- Energy reliance is decreased fundamentally since watering should be possible using a low-weight zone.
- Productivity is increased between 20 and 90% naturally.
- The products ripened sooner.
- Cultural exchanges can be executed more effectively. Working for the demand is reduced.
- Water loss is negligible, and water saving increased by 50%.
- The drip irrigation system can apply the manure and disinfectant effectively.
- Ground and soil erosion loss is prevented.
- It is effectively utilized on downsizing land.
- This system makes less ground hardening.
- Groundwater is increased effectively, empowering plants to develop all the more healthily. You will profit by high caliber, productive, and institutionalized item.
- The development of the plant is subsequently institutionalized, giving persistent field limit because of short and discrete water exchanges.
- It guarantees productive water utilization, regardless of whether the assets contain a high level of salt. (Incidentally—the misfortune proportion for plants decreases).
- This makes an advantage from a 60% reserve funds in manure and purification costs (contrasted with traditional water system frameworks).
- Loss of water never occurs since surface stream and surge to the profundity are both averted.

- It gives conceivable watering outcomes even in fields that don't have enough regular water assets.
- It provides a one-time harvesting opportunity since equivalent water and fertilizer sharing between plants creates extraordinary advantages.
- Plants need a more noteworthy starting point to get water in connected using old style water system frameworks.
- Adding points of interest to dribble water system strategy by implementing robotization increment proficiency, lessen work cost, increment efficiency, and so on.

1.21   Time-Based System

The water required and the flow rates of water are determined by the time of activity. In the system, time is the premise of the water system. The span of individual valves must be maintained in controller alongside the starting time of framework. Likewise, the controller clock is to be set with the present day and time (Rajakumar et al. 2005). The system is having the drawback of no feedback to improve the operation efficiency of the system and less precision in the valve control.

1.22   Volume-Based System

In volume-based framework, the present measure of water can be connected in the field sections by utilizing programmed volume controlled metering valves (Rajakumar et al. 2005). The main problem with the volume-based system is the high cost for implementation, complex system design, and the amount of water to be supplied, which is by the use of keypad only.

1.23   Open-Loop Systems

The water supply level is decided by the operator in the open loop system. The information is programmed in the controller, and the controller acts to supply the water based on the schedule given (Rajakumar et al. 2005). The failure of the system is because of not been to react naturally for the change in the earth conditions.

1.24   Closed-Loop Systems

The closed-loop system is more controlled and gives higher efficiency. The sensors provide feedback to the system. The control system takes the decision of level and the time of water supply to the plants (Rajakumar et al. 2005). The drawback of the system is that the controller must be reprogrammed for different types of cultivation, control status, and other required data not displayed to the user, and data acquisition is not performed well.

1.25 Existing Irrigation Systems

1. The interest in new water sparing methods in the water system is expanding quickly at present. To create "more yield per drop," producers in (semi) parched locales as of now investigate water system strategies in the range from utilizing less crisp water.
2. A short time later, sensors for soil dampness and sprinkler valve controllers are started for use in site-explicit water system computerization (Valente et al. 2007).
3. As far as conventions are concerned, infrared, wireless personal area networks, Bluetooth, wireless local area networks have been put to various utilities to execute remote sensors in accuracy horticulture.
4. In most recent two decades, with the improvement of remote innovations, a few inquiries were concentrated on self-governing water system with sensors in rural frameworks.

1.26 Proposed Irrigation System Based on Closed-Loop Feedback System

Constant input is the application of water system dependent on the real compelling interest of the plant itself, plant root zone successfully mirroring every single ecological factor following up on the plant. Working inside for the different controlled parameters and the plant itself decides the level of water system required. Various sensors, namely, moisture and temperature sensors, control the water system planning. These sensors give input to the controller to monitor its activity.

Use of a remote sensor arranges for minimal effort real-time remote input controlled water system arrangement and constant observing of water substance of soil (Vories et al. 2003). Battery controlled remote obtaining station sends information to control station with the end goal of controlling valves for the water system, and the control can be in two-mode activities: Automatic and manual which utilizes Global System for Mobile Communications innovation (Zhao et al. 2007). The proposed system has the advantage of having a nozzle that can be controlled with greater efficiency, reducing the labor cost, and improving the effectiveness of the water supply. Without using much energy, the plant gets its water, and this promotes sufficient growth. The user knows the status of the valves, temperature and moisture level, and the power status. Depending on the information from the base station unit (BSU), the development procedure can be changed.

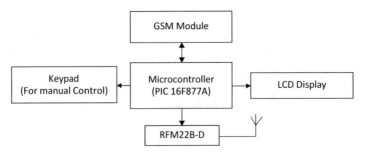

Fig. 1.27 Block diagram of the control station

1.26.1 Control Station

The microswitches are used to get the input from the user. Then commands are sent from control station unit (CSU) to the base station. Based on control signal the base station will be operated with a valve such as ON/OFF. There are six microswitches used in control station. Three microswitches are used to ON/OFF the valve1, valve2, and valve3. Confirmation switch is used to confirm the signals before sending from the control station to the base station via RFM22B.

GSM module (SIM300 v1.01) is interfaced to CSU to update the status via SMS. Keywords such as STAT and V1 ON are used as control commands. Figure 1.27 shows a block diagram of the control station unit.

1.26.2 Base Station

Base station unit, valve unit, and sensor unit are the units available in the field. The base station will receive commands from a control station via RFM22B. Figure 1.28 shows a block diagram of the base station. The base station will control the valves and the motor. Moisture content and temperature are the parameters to be measured in the field in which temperature is measured by using the inbuilt temperature sensor in RFM22B, and capacitive type sensor is used for measuring moisture content.

1.26.3 Hardware Description

This station includes hardware such as

- 16F877A- PIC microcontroller.
- RFM22B-Digital wireless transceiver.
- SIM300v7.03-GSM module.
- HD44780-LCD.

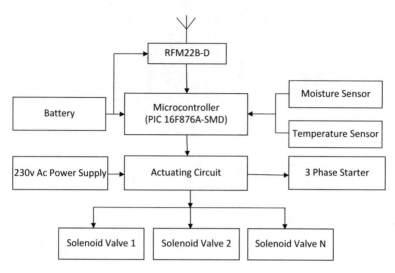

Fig. 1.28 Block diagram of the base station

These were the major hardware that contributed to the effective functioning of our application.

1.26.4 Circuit Diagram Explanation

Figure 1.29 refers to control station circuit, and circuit explanation is given below:

- GSM module was interfaced to the microcontroller via MAX232 using serial communication for data exchange.
- RX/TX pins were used for receiving and transmitting data, respectively.
- MAX 232 was DIP IC for interfacing two modules so that no low-voltage problem occurs and at the same time data error is eliminated.
- RFM22B (digital wireless transceiver) is interfaced to the microcontroller via SPI protocol using three wires, namely SDO, SDI, and SCK.
- Microswitches are used for keypad control mode.
- LCD is interfaced in 4bit mode, and RS, RW, and EN were the pins used for LCD control.
- nSEL is the pin used to select the RFM22B to work for transmission or reception mode.
- nIRQ is the pin used to check whether the transceiver is attempting to interrupt.
- TXEN and RXEN are the pins used to select transceiver to operate as transmitter/receiver.

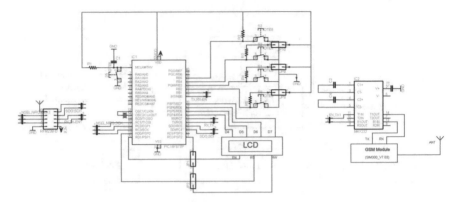

Fig. 1.29 Control station circuit diagram

1.26.5 Working

- Commands to the base station are sent through only control station via RFM22B (digital wireless transceiver).
- Commands are given through keypads and SMS to control station, and then RFM22B transmits it to the base station.
- The base station will receive it and respond to the control station based on command and process the digital data, and it will transmit the processed data back to the control station.
- If the command is a control signal, then it controls the relays and updates the status in control station.
- GSM module is used to send SMS/to receive the SMS via PIC microcontroller.
- So SIM300v7.03 (GSM module) is entirely under the control of 16f877a microcontroller.
- Status can be updated via LCD and SMS to the user.
- This makes the application as user-friendly for control.

1.26.6 Base Station Circuit Hardware Description

- Current transformer (CT-2500 T).
- 16F877A- PIC microcontroller.
- RFM22B-Digital wireless transceiver.
- Soil moisture sensor.
- Solenoid valves.
- Battery panel and battery.

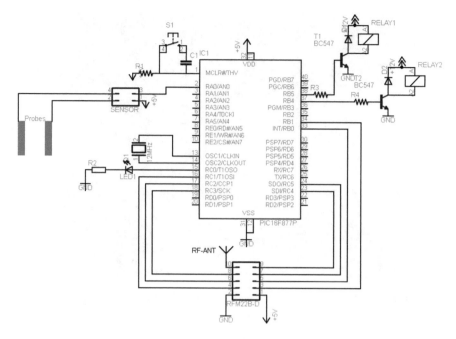

Fig. 1.30 Base station

1.26.7   Circuit Diagram Explanation

Figure 1.30 refers to base station circuit:

- RFM22B (digital wireless transceiver) is interfaced to the microcontroller via SPI protocol using three wires (SDO, SDI, SCK).
- Inbuilt ADC of the microcontroller is used for analog to digital conversion.
- Actuating circuit consists of relay drivers and solenoid valves.
- Relays are driven using BC547 NPN transistor.

1.26.8   Working

- Whenever the base station receives the signal from the control station, it will process the digital data from ADC.
- The inbuilt temperature sensor in RFM22B is used here to get temperature for the comparison process.
- 8bit digital output from RFM22B is compared to the digital output of ADC in microcontroller and status is updated in the base station, and it is transmitted to the control station.

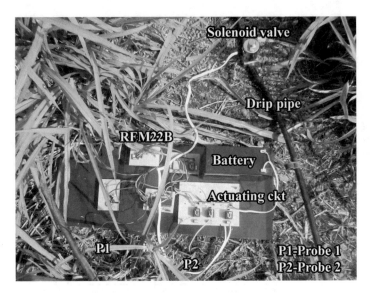

Fig. 1.31 System's application around the plant with the sensor unit

- If the base station receives the control commands, then it will control the respective relay and update the status and transmit to the control station.

1.26.9   Experimental Study

The base station unit is located in the field including sensors, solenoid valve, and a transceiver. Capacitive type sensor is used to measure the moisture content and a hacksaw blade used as a capacitor plate and soil acting as a dielectric medium. Square wave from Astable Multivibrator was given as an input to Probe 1, and probe 2 is connected to PORTA pin A0 of 16f877A PIC microcontroller. A Solenoid valve is connected in between the pipes, to control the water flow to the plant. A battery powers total circuit except for solenoid valve because here we are using 230 V, 6 W solenoid valve for demo purpose, so 230v ac supply is required. Figure 1.31 shows system application around a plant representing the base station unit.

The solenoid valve is connected in between the two drip pipes, so the water flow through the pipe is controlled by a valve. P1 and P2 are the probes acting as a capacitor plate, and here we used a hacksaw blade as a probe because it is like plate so it is suitable to act as a capacitor plate and distance between the two probes are around 15 cm. RFM22B is a digital wireless transceiver used here to transmit data from the field and to receive commands from the CSU. Actuating circuit is used to control the solenoid valve, and it includes relays and its driving circuit.

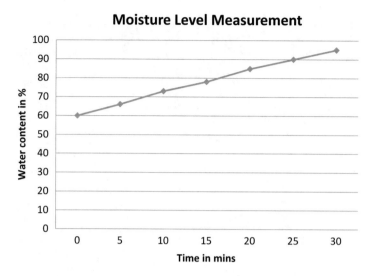

Fig. 1.32 Moisture level measurement

The entire system was implemented in agricultural land and tested to our best knowledge. The base station unit received the information during the irrigation time from 3.00 pm to 3.30 pm. The data was recorded every 5 min once. The graph is plotted for recorded data, which is represented by a line chart and given in Fig. 1.32. Figure 1.32 shows moisture level measurement.

When the moisture reaches 95%, the actuating circuit turns off the solenoid valve. For a demonstration, only one solenoid valve is used. Figure 1.33 shows soil reaching 90% of moisture near the BSU.

1.26.10   Starter Connection

Three-phase starter is used for agricultural pump sets. Push buttons are used to turn ON/OFF the motor, in which ON (Green) button is in NO position, so relay associated to that is connected in NO, and OFF (Red) button is in NC position. Figure 1.34 shows relay connection to 3 phase starter.

1.26.11   Control Station at Home

Control station unit is located in the home and from there the BSU can be controlled, and the status of BSU can be updated in the control station unit. Figure 1.35 shows CSU located at home.

Fig. 1.33 After the moisture, level reaches 95%

Fig. 1.34 Phase starter with relay circuit

All analog values from the sensor are converted to digital data using inbuilt ADC in PIC microcontroller of BSU, and those digital values are transmitted by BSU and received by CSU and processed in CSU. Commands are sent to BSU through CSU and responses are received in CSU.

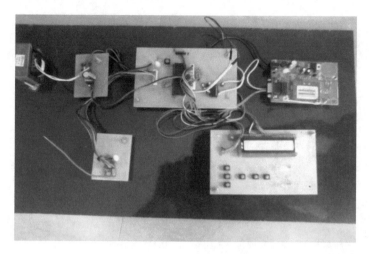

Fig. 1.35 Control station unit

This system was effectively implemented and tested in agricultural land and focused on future implementation, including PC interface for data storing and analyzing for seasonal changes to adopt suitable cultivation.

Smart Agriculture Application

Agriculture is a significant economic influencer in Botswana. Although this is true, the support of the agricultural industry to Botswana's gross domestic product dropped significantly to 2.1% in 2018, as showed by a report published by Market and Research [http://apanews.net/en/news/botswana-agric-sectors-contribution-to-gdp-declines-report]. It has been observed that agriculture sector supported nearly 2.4% to Botswana's GDP before 2017. The research conducted by the Ireland-based research group studied that 42% of the southern African country's population inhabits the rural areas as mentioned in the data of the World Bank. It has been revealed from the data that nearly 70% of rural households in Botswana rely on the growth of crops for consumption. The statistic showed that crop production contributes 20%, whereas livestock contributed 80% of the income in the agricultural [http://apanews.net/en/news/botswana-agric-sectors-contribution-to-gdp-declines-report]. With this said, initiatives must be formed to boost the agricultural sector. The study reveals that the least performing area in agriculture is crop production in comparison to livestock farming. The primary reason is crop farming has a higher sensitivity than livestock farming. Relevant knowledge has to be available and practiced for improved efficiency. The use of an agricultural app is proposed, as shown in Fig. 1.36, to readily provide information on different crops in Botswana, their compatibility with an area, what amount of water and nutrients they need, etc.

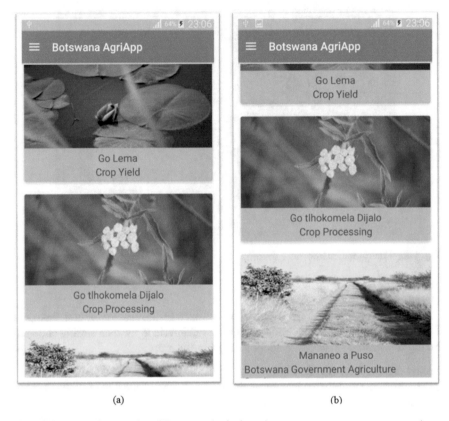

(a)   (b)

Fig. 1.36 Screen shot samples of Botswana Agriculture App

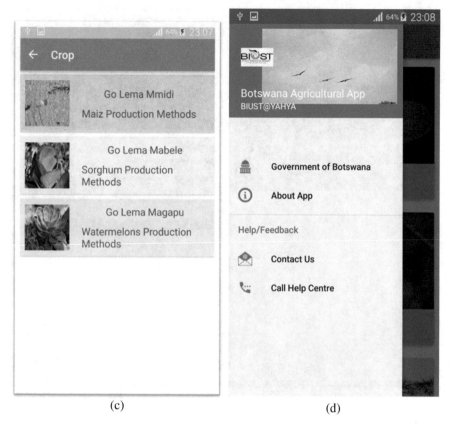

(c)   (d)

Fig. 1.36 (continued)

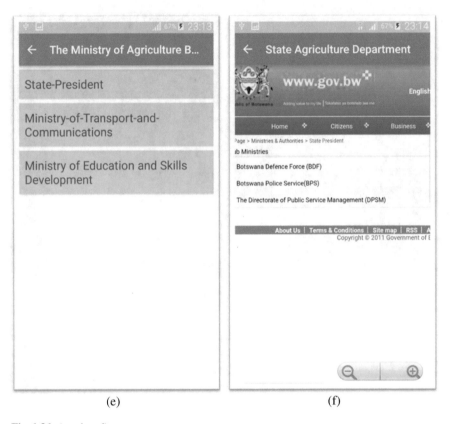

(e)   (f)

Fig. 1.36 (continued)

References

Abd El-kader SM, El-Basioni BMM (2013) Precision farming in Egypt using wireless sensor network technology. Egyptian Informatics Journal 14:221–233

Aiello G, Giovinoa I, Valloneb M, Cataniab P, Argentoa A (2018) A decision support system based on multisensor data fusion for sustainable greenhouse management. Journal of Cleaner Production 172:4057–4065

Bewley JM, Russell RA, Dolecheck KA, Borchers MR (2015) Precision what have we learned. In: Halachmi I (ed) Precision Livestock Farming Applications-Making Sense of Sensors to Support Farm Management. Wageningen Academic Publishers, Wageningen, pp 13–24

Bhuyan B, Sarma HK, Sarma N (2014) A survey on middleware for wireless sensor networks. J Wireless Network Communication 4(1):7–17

Brown B, Nuberg I, Llewellyn R (2018) Stepwise frameworks for understanding the utilisation of conservation agriculture in Africa. Agr Syst 153:11–22

Cao CM, Xia P, Zhu ZQ (2007) Application of wireless data transmission to the automatic control of water saving irrigation. Transactions on Chinese Society of Agricultural Engineering 21(2):127–130

Capello F, Toja M, Trapani, N (2016) A real-time monitoring service based on industrial internet of things to manage agrifood logistics. 6th International Conference on Information Systems, Logistics and Supply Chain, pp. 1–8

Chaudhary DD, Nayse SP, Waghmare LM (2011) Application of wireless sensor networks for greenhouse parameter control in precision agriculture. International Journal of Wireless & Mobile Networks (IJWMN) 3:140–149

Chi M, Plaza A, Benediktsson JA, Sun Z, Shen J, Zhu Y (2016) Big data for remote sensing: challenges and opportunities. Proc IEEE 104(11):2207–2219

Chong C, Kumar SP (2003) Sensor networks: evolution, opportunities, and challenges. Proc IEEE 91(8):1247–1256

Collinson MP (2000). A History of Farming Systems Research Cabi Press, Oxford

Danuri MSNM, Shahibi MS (2015). The Development of Farm Management Information System for Smallholder Farmers in Malaysia. Proceeding of IC-ITS 2015, International Conference on Information Technology & Society 8–9 June 2015, Kuala Lumpur, Malaysia

Fang S, Da Xu L, Zhu Y, Ahati J, Pei H, Yan J, Liu Z (2014) An integrated system for regional environmental Monitoring and management based on internet of things. IEEE Transactions on Industrial Informatics 10(2):1596–1605

Fountas S, Carli G, Sørensen CG, Tsiropoulos Z, Cavalaris C, Vatsanidou A, Liakos B, Canavari M, Wiebensohn J, Tisserye B (2015) Farm management information systems: current situation and future perspectives. Comput Electron Agric 115:40–50

Fourati MA, Chebbi W, Kamoun A (2014) Development of a web-based weather station for irrigation scheduling. Third IEEE International Colloquium in Information Science and Technology (CIST), Tetouan, pp. 37–42

Fox L, Pimhidzai O (2013) Household non-farm enterprises and structural transformation: evidence from Uganda. Paper presented at the UNU-WIDER Conference on Inclusive Growth in Africa, Helsinki

Gaire R, Lefort L, Compton M, Falzon G, Lamb D, Taylor K (2013) Demonstration: semantic web enabled smart farm with GSN. http://ceur-ws.org/Vol-1035/iswc2013_demo_11.pdf

Gutiérrez PA, López-Granados F, Peña-Barragán JM, Jurado-Expósito M, Martíneza C (2008) Logistic regression product-unit neural networks for mapping Ridolfia segetum infestations in sunflower crop using multitemporal remote sensed data. Computers and Electronics in Agriculture. 64:293–306. https://doi.org/10.1016/j.compag.2008.06.001

Gutiérrez J, Villa-Medina JF, Nieto-Garibay A, Porta-Gándara MÁ (2014) Automated irrigation system using a wireless sensor network and GPRS module. IEEE Trans Instrum Meas 63(1):166–176

Haggblade S, Hazell P, Reardon T (2010) The rural non-farm economy: prospects for growth and poverty reduction. World Dev 38(10):1429–1441

Hashim N, Mazlan S, Aziz MA, Salleh A, Jaafar A, Mohamad N (2015) Agriculture monitoring system: a study. Jurnal Teknologi 77(1):53–59

Husemann C, Novkovic N (2014) Farm management information systems: A case study on a German multifunctional farm. Economics of Agriculture 61(2):175293

Kacira M, Sase S, Okushima L, Ling PP (2005) Plant response-based sensing for control strategies in sustainable greenhouse production. Journal of Agricultural Meteorology 61(1):15e22

Karim F, Karim F, frihida A (2017) Monitoring system using web of things in precision agriculture. Procedia Computer Science 110:402–409

Kassam A, Friedrich T, Derpsch R, Kienzle J (2015) Overview of the worldwide spread of conservation agriculture. Field Actions Science Reports [Online] 8

Keshtgari M, Deljoo A (January 2012) A wireless sensor network solution for precision agriculture based on ZigBee technology. Wirel Sens Netw 4:25–30

Kittas C, Bartzanas T, Jaffrin A (2003) Temperature gradients in a partially shaded large greenhouse equipped with evaporative cooling pads. Biosyst Eng 85(1):87–94

Kodali RK, Rawat N, Boppana L (2014) WSN sensors for precision agriculture. 2014 IEEE Region 10 Symposium, Kuala Lumpur, pp. 651–656

Kok R, Gauthier L (1986) Development of a prototype farm information management system. Computers and Electronics in Agriculture Vol 1:125–141

Laney D (2001) 3D data management: controlling data volume, velocity and variety. META Group Research Note, 6

Lanjouw J, Lanjouw P (2001) The rural non-farm sector: issues and evidence from developing countries. Agric Econ 26(1):1–23

Lee J, Park GL, Kang MJ, Kwak HY, Lee SJ, Han J (2012) Middleware integration for ubiquitous sensor networks in agriculture. In: Murgante B et al (eds) Computational science and its applications – ICCSA 2012. ICCSA 2012. Lecture notes in computer science, vol 7335. Springer, Berlin, Heidelberg

Li M, Chen G, Zhu Z (2013) Information service system of agriculture IoT. Automatika J control, Meas Electron Comput Commun 54:415–426

Lokers R, Knapen R, Janssen S, Randen YV, Jansen J (2016) Analysis of big data technologies for use in agro-environmental science. Environ Model Softw 84:494–504

Luan Q, Fang X, Ye C, Liu Y (2015) An integrated service system for agricultural drought monitoring and forecasting and irrigation amount forecasting. 23rd IEEE International Conference on Geoinformatics, pp. 1–7

Market and Research (2018) Botswana agric sector's contribution to GDP declines—Report, [Online]. Available: http://apanews.net/en/news/botswana-agric-sectors-contribution-to-gdp-declines-report

Martin MA, Islam MM (2012) Overview of wireless sensor network. IntechOpen, London, pp 1–23

Michael O, Gregory O (2017) Modelling the smart farm. Information Processing in Agriculture. 4. https://doi.org/10.1016/j.inpa.2017.05.001

Milovanovic S (2014) The role and potential of information technology in agricultural improvement. Ekonomika Poljoprivrede 61(2):471

Mohanraj I, Ashokumar K, Naren J (2016) Field monitoring and automation using IOT in agriculture domain. Procedia Computer Science 93:931–939

Mohd Kassim MR, Mat I, Harun AN (2014) Wireless Sensor Network in precision agriculture application. International Conference on Computer, Information and Telecommunication Systems (CITS), Jeju, pp. 1–5, 2014

Muangprathub J, Boonnam N, Kajornkasirat S, Lekbangpong N, Wanichsombat A, Nillaor P (2019) IoT and agriculture data analysis for smart farm. Comput Electron Agric 156:467–474

Mupangwa W, Mutenje M, Thierfelder C, Nyagumbo I (2016) Are conservation agriculture (CA) systems productive and profitable options for smallholder farmers in different agro-ecoregions of Zimbabwe? Renewable Agric Food Syst pp 32(1):87–103

Nack F (2008–2009) An overview on wireless sensor networks. [Online]. Available: https://www.mi.fu-berlin.de/inf/groups/ag-tech/teaching/2008-09_WS/S_19565_Proseminar_Technische_Informatik/nack09verview.pdf

Nagler P, Naudé W (2017) Non-farm entrepreneurship in rural sub-Saharan Africa: new empirical evidence. Food Policy 67:175–191

Ndah HT, Schuler J, Uthes S, Zander P, Traore K, Gama MS, Nyagumbo I, Triomphe B, Sieber S, Corbeels M (2014) Adoption potential of conservation agriculture practices in sub-Saharan Africa: results from five case studies. Environ Manag 53:620–635

Ndzi DL, Harun A, Ramli FM, Kamarudin ML, Zakaria A, Shakaff AYM, Farook RS (2014) Wireless sensor network coverage measurement and planning in mixed crop farming. Comput Electron Agric 105:83–94

Nishina H (2015) Development of speaking plant approach technique for intelligent greenhouse. Agriculture and Agricultural Science Procedia 3:9–13

Norman DW (1995) The farming systems approach to development and appropriate technology generation (No. 10). Food & Agriculture Org

O'Hare GMP, Muldoon C, O'Grady MJ, Collier RW, Murdoch O, Carr D (2012) Sensor web interaction. International Journal on Artificial Intelligence Tools 21(2):1240006

Pang Z, Chen Q, Han W, Zheng L (2015) Value-centric design of the Internet-of Things Solution for food supply chain: value creation. Sensor Portfolio and Information Fusion Inform Syst Front Vol 17:289–319

Paraforos D, Vassiliadis V, Kortenbruck D, Stamkopoulos K, Ziogas V, Sapounas AA, Griepentrog HW (2016) A farm management information system using future internet technologies. IFAC-PapersOnLine 49(16):324–329

Rajakumar D, Ramah K, Rathika S, Thiyagarajan G (2005) Automation in micro irrigation. Technology Innovation Management and Entrepreneurship Information Service, New Delhi

Reardon T, Berdegue J, Barrett C, Stamoulis K (2006) Household income diversification into rural non-farm activities. Johns Hopkins University Press, Baltimore

Roham VS, Pawar GA, Patil AS, Rupnar PR (2015) Smart farm using wireless sensor network. International Journal of Computer Applications (0975–8887), National Conference on Advances in Computing NCAC 2015(\), pp. 8–11

Savale O, Managave A, Ambekar D, Sathe S (2015) Internet of things in precision agriculture using wireless sensor networks. International Journal of Advanced Engineering & Innovative Technology (IJAEIT) 2(3)

Shen WS, Liu G, Su Z, Su R, Zhang Y (2016) Design and implementation of livestock house environmental perception system based on wireless sensor networks. International Journal of Smart Home 10:69–78

Shock CC, David RJ, Shock CA, Kimberling CA (1999) Innovative, automatic, low-cost reading of Watermark soil moisture sensors. Proceedings of the International Irrigation Show, pp. 147–152

Start D (2001) The rise and fall of the rural non-farm economy: poverty impacts and policy options. Dev Policy Rev 19(4):491–505

Talavera JM, Tobón LE, Gómez JA, Culman MA, Aranda JM, Parra DT, Quiroz LA, Hoyos A, Garreta LE (2017) Review of IoT applications in agro-industrial and environmental fields. Computers and Electronics in Agriculture 142(Part A):283–297

Thierfelder C, Bunderson W, Mupangwa W (2015) Evidence and lessons learned from long-term on-farm research on conservation agriculture Systems in Communities in Malawi and Zimbabwe. Environments 2:317–337

Thierfelder C, Matemba-Mutasa R, Bunderson WT, Mutenje M, Nyagumbo I, Mupangwa W (2016) Evaluating manual conservation agriculture systems in southern Africa. Agric Ecosyst Environ 222:112–124

Thiombiano L, Meshack M (2009) Scaling-up conservation agriculture in Africa: strategy and approaches. FAO, Addis Ababa

Thompson SC (1976) Canfarm—a farm management information systems. Agric Administr Vol 3:181–192

Tzounis A, Katsoulas N, Bartzanas T, Kittas C (2017) Internet of things in agriculture, recent advances and future challenges. Biosyst Eng 164:31–48

Valente R, Morais C, Serodio P, Mestre S, Pinto S, Cabral M (2007) A ZigBee sensor element for distributed monitoring of soil parameters in environmental monitoring, IEEE Conference on Sensors, pp. 135–138

Verdouw CN, Beulens AJM, van der Vorst JGAJ (2013) Virtualisation of floricultural supply chains: a review from an internet of things perspective. Comput Electron Agric 99:160–175

Vories ED, Glover RE, Bryant KJ, Tacker PL (2003) Estimating the cost of delaying irrigation for mid-south cotton on clay soil, Proceedings of the 2003 Beltwide Cotton Conference National Cotton Council Memphis, pp. 656–661

Wang MM, Cao JN, Li J, Dasi SK (2008) Middleware for wireless sensor networks: a survey. J Comput Sci Technol 23(3):305–306

Wark T, Corke P, Sikka P, Klingbeil L, Guo Y, Crossman C et al (2007) Transforming agriculture through pervasive wireless sensor networks. IEEE Pervasive Computing 6(2):50–57

Whitfield S, Dixon JL, Mulenga BP, Ngoma H (2015) Conceptualising farming systems for agricultural development research: cases from eastern and southern Africa. Agr Syst 133:54–62

Whitman EC (2005) SOSUS: The Secret Weapon of Undersea Surveillance. Undersea Warfare 7(2)

Yang DT, Zhu X (2013) Modernization of agriculture and long-term growth. Journal of Monetary Economics 60(3):367–382

Zhao YD, Bai CX, Zhao B (2007) An automatic control system of precision irrigation for City greenbelt, Proceedings of Second IEEE Conference on Industrial Electronics and Applications

Zhao J-c, Zhang J-f, Yu F, Guo J-x. The study and application of the IOT technology in agriculture. 3rd International Conference on Computer Science and Information Technology, Chengdu, pp. 462–465, 2010

Chapter 2
Livestock: Monitoring Foot and Mouth Disease Using Wireless Sensor Networks

Abstract This chapter presents a system to detect foot and mouth disease as early as possible in a herd of cattle using wireless sensor networks. The system combines animal behavior and sensor values to determine the status of the cattle in terms of foot and mouth disease. The system combines animal behavior and sensor values to determine the status of the cattle in terms of foot and mouth disease. The system first measures an average response of a cow under normal circumstances with the focus being on the measurements of body temperature; distance covered; and feeding rate every 2 h. Data regarding the state of the cow is sent at certain time intervals to the farmers using ZigBee via a gateway. This chapter provides results of the trials performed on the project. A signal is acquired from a cow sensor node and transmitted to the Gateway which is interfaced to a computer running LABVIEW software as the graphical user interface of the system. The trials were done both for negatively and positively diagnosed cows.

2.1 Introduction

Certain parts of a cow's body possess sure signs and symptoms when foot and mouth disease (FMD) infect the cow. The affected regions are usually the tongue, hooves, and teats. Since a cow develops a high fever during the incubation period of the FMD virus, it is essential to monitor the body temperature of the cow, which is a parameter to serve as an alarm for fever. An infected cow is also known not to feed well due to blisters that develop on the tongue, making it hard to feed. By monitoring the food intake of the cow, it will be possible to tell a possible FMD case. The blisters that appear on the hooves of the feet will have to be detected with sensors as well. It is vital for the personnel monitoring the cattle to have real-time data so that they can perform analyses on the received data and make decisions upon suspicious events. The data should be sent back to the monitoring station using radio transceivers. The cows are being monitored for their health status in terms of FMD, which means an immediate action must be taken upon the detection of the disease. One such activity is following the cow into the grazing land to isolate it. This means the precise location of the cow must be known at all times for this reason; the project uses a Global Positioning System (GPS) tracker to determine the site of a cow. The tracking device provides the geographical coordinates of the location of the cow in

© Springer Nature Switzerland AG 2020
A. Yahya, *Emerging Technologies in Agriculture, Livestock, and Climate*,
https://doi.org/10.1007/978-3-030-33487-1_2

real time. The system is designed to have two parties being able to access the telemetric data. These parties are the cattle owner and veterinary officers. Since FMD-infected cattle must be quarantined with an immediate effect, the veterinary officers must also be notified if FMD is caught in the local animals. This also serves useful if the farmer hesitates to take action about the detected FMD.

Figure 2.1 shows the system is connected to the Cloud to enable monitoring from all corners of the world. The cattle have sensors attached to them as well as GPS tracking devices which communicate with the respective satellites. The cattle farming site is equipped with a gateway that fetches data from the cattle and passes it on to the users via the Cloud. The Cloud access methods, in this case, are General Packet Radio Service (GPRS) and Very Small Aperture Terminal (VSAT). GPRS will be applicable in instances where the Global System covers the cattle farming site for Mobile (GSM) communications cellular networks infrastructure. VSAT is brought into use in case there is no GSM coverage. The radio being employed is to operate in the Institute of Electrical and Electronics Engineers (IEEE) 802.15.4 Industrial, Scientific and Medical (ISM) frequency bands. The reason for choosing this band is because there is no license required to operate in it. Also, the devices running in this band have a feature of interoperability which allows them to mitigate interference. The remote end system and users shown in Fig. 2.1 are the Veterinary officers and the other members authorized accessing the telemetric data. The project also gives a profile to each cow. The profile information is inclusive of the name of cow, age, instantaneous FMD status as diagnosed by the employed sensors, as well as immediate location.

2.2   Definition and Epidemiology of FMD

According to Jamal and Belsham (2013), FMD is a disease that affects cloven-hoofed animals such as cattle, pigs, small stock, and a variety of wild animals like the African buffalo. The authors carry on stating that in most parts of the world where there is pastoral farming, FMD has prevailed at one point in time. FMD is caused by a virus strain referred to by its scientific name as *Aphthovirus Picornaviridae*. The FMD virus is said to have seven variants or types being A, C, O, Asia-1, Southern African Territories 1 (SAT 1), SAT 2, and SAT 3. These seven types also have subtypes within them, which are difficult to classify due to their random patterns. Donaldson (1993) has provided an insight into the epidemiology of FMD across the world. According to the author, in South America types A, O, and C exist with O and A being the most prevalent. Asia experiences the types in order from the most pervasive being O, A, Asia-1, and C. The author carries on stating that the northern parts of Africa see type O as the predominant. The west, east, and central parts of Africa are said to experience types SAT 1 and 2, O, A, and C. Southern Africa are known only for the SAT series of the virus strains. Thomson (1995) mentioned that the African buffalo is the primary source of FMD towards livestock. The author carried on stating that the northern parts of Botswana have been the ones suffering from FMD attacks mostly within the country. The probable

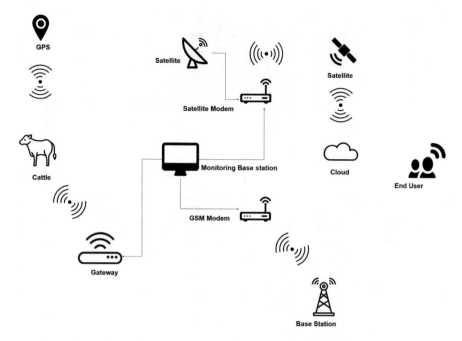

Fig. 2.1 The cattle FMD monitoring system

reason for this is because Botswana in the north shares a border with Zimbabwe, Zambia, and Namibia. The area where these three countries converge is part of the focus of FMD in the trio with contributions from southeast Angola. The same area is characterized by wildlife inclusive of the African buffalo, which is the primary source of FMD. This is, therefore, a likely reason for the significant prevalence of FMD in the northern parts of Botswana. Figure 2.2 is a map of Botswana showing FMD zones. According to the plan, the parts of the north are the only ones where there is FMD vaccination because the disease is more rampant in the region.

2.3 History of FMD

Jamal and Belsham (2013) have stated that FMD was first discovered in 1514 in Venice, Italy. The FMD virus type O was found in France while type A in Germany. The discoveries of these two types were followed by type C and then SAT classes in South Africa. The type Asia-1 was first found in Pakistan in 1954 from a water buffalo. Pattnaik et al. (2012) have mentioned that FMD in the Southern African region was first discovered in South Africa in the year 1892 and it later led to an endemic case according to Thomson (1995) during the era 1896–1905 in Southern Africa. The FMD then disappeared and resurfaced in March 1931. Thomson (1995) carried on stating that the disease has been terrorizing the southern African economy since

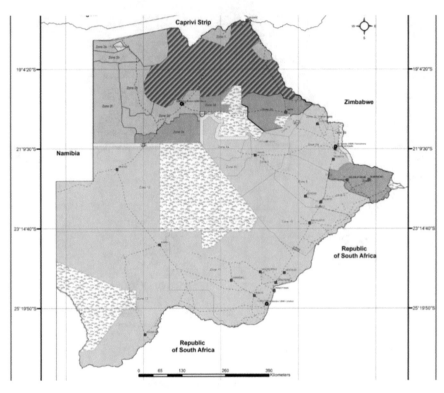

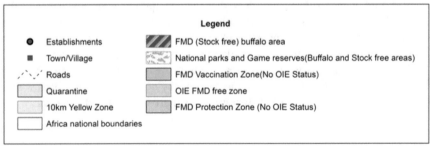

Fig. 2.2 Map of Botswana showing FMD stricken zones

it resurfaced in 1931 with frequent outbreaks. Botswana was no exception in being victimized by FMD. Since 1931, Botswana has had FMD outbreaks in every decade. The period 1971–1980 is said to have led to the most massive FMD outbreaks in the southern African region. The FMD outbreaks during this period are attributed to the political unrests that took place in Angola and Mozambique. The political uprisings in the two countries meant there was no control of FMD, which led to its spread to the neighboring countries.

2.4 Transmission and Pathogenesis of FMD

The transfer of FMD was described by Aftosa (2014) for most animals of domestic and wild types. As the author begins, they stated that the FMD virus could be carried by most secretions and excretions of infected animals. Such secretions and excretions include saliva, exhaled air, semen, milk, feces, and urine. The author carried on mentioning that vesicles vulnerable to FMD also produce fluids that may take the FMD virus. Such may be the amniotic fluid or fetuses due to abortions in sheep. The FMD virus is said to have its routes into an animal through inhalation, ingestion, skin abrasions or bruises and mucous membranes. Still, according to the author, cattle are most prone to being infected by an aerosolized virus. The FMD virus can also spread over water too long distances over 250 km. Aftosa (2014) stated that the FMD virus could survive under different conditions such as cold, where it may live up to 1 year. It is also supported by organic matter and sunlight with life durations up to 3 months. Meat and milk can also help FMD viruses for long times. According to Ashford (2019; http://www.msdvetmanual.com/generalized-conditions/foot-and-mouth-disease/overview-of-foot-and-mouth-disease), FMD can be transmitted by mechanical vectors as well. The mechanical vectors such as human beings, vehicles, and animals aid in the transmission of the disease. On human beings, the virus may be carried on the skin or clothes, and as the human being moves, that is how they spread the FMD virus. Aftosa (2014) also stated that humans could also carry the FMD virus through the nasal passage for up to 28 h. The virus can also be taken by vehicles as they move from one region to another. Animals such as dogs, horses, and pigs also transmit the virus with pigs being victims of FMD as well.

2.5 Signs and Symptoms of FMD

The Cattle Site (2014) mentioned that symptoms of FMD include fever, blisters in the mouth, feet, and sometimes teats, reduced milk production, loss of weight and appetite, quivering lips and frothing of mouth and lameness. Cabi (2019, https://www.cabi.org/isc/datasheet/82822) stated that the FMD incubation period is 2–11 days. The incubation period is dependent on the virus strain and dose. The author continues to report that during the incubation period, the infected cow develops fever, and cysts develop in the tongue and buccal mucosa, the coronary band, and heels of the feet, and on the nipple. Tho (2017) highlighted there is a treatment that can be given to the infected cattle to prevent secondary infections. This treatment also enhances the ability of the cow to recover from FMD. With an early detection mechanism in place, the disease will be detected before the arrival of secondary infections which will give farmers or veterinary officers enough time to treat the cattle as prevention of secondary infections and also to enhance the recovery of a cow.

Fig. 2.3 Cow ear tags

2.6 Methods of Control of FMD

2.6.1 Livestock Identification and Traceability Systems (LITSs)

Prinsloo et al. (2017) described a method of reducing the spread of FMD primarily based on traceability systems. The authors took a case study in Namibia. The traceability system keeps track of an animal alive or dead or converted into a product. As an explanation, the authors carry on stating that to trace live cattle, documents working in conjunction with cattle ear tags are used to update the information regarding the location of the cows as they are moved from one place to another. This may be from one farm to another, from a farm to a market place, etc. In moving the animals, permission must be requested from veterinary services to advance the cow. The request for the movement of the cows is made through the use of documents identifying the cows. Upon the initiation of the action, the ear tags, as shown in Fig. 2.3, and veterinary permits are checked to see if they match. The same procedure of checking the ear tags against the documents is done at the destination. That way a cow's origin can be traced if it is found to possess FMD. The authors also explained how to locate finished products. Important information is posted on the containers of meat products. The information may be the original country of the animal, sex, age, places of birth, and slaughtering and the meat processing site. With this system, finished products can be traced back from consumers to producers via retailers, distributors, and meat processing plants. That will allow for an easy determination of the source of an outbreak if it occurs. As they continue the authors highlighted on the control of FMD in Namibia in incidents of outbreaks by describing a system referred to as Namibia Livestock Identification and Traceability System (NamLITS) which is used to determine the areas being affected by FMD outbreaks. Once the affected areas are identified, quarantines are used as a measure of preventing further spread. The constraint of this technique is that the FMD is detected at advanced stages when little can be done to treat the disease and possibly when the range has gone too far due to the late detection.

2.6.2   Vaccination

Pattnaik et al. (2012) stated that the first FMD vaccine was developed in 1926 and was followed in 1937 by a vaccine to be actively used against FMD. Derah and Mokopasetso (2005) highlighted vaccination as a control measure of FMD in Botswana. In Botswana, a governmental organization referred to as Botswana Vaccines Institute (BVI) is responsible for producing FMD vaccines. The vaccines from BVI target the types SAT 1, SAT 2, SAT 3, O, and A. BVI (2019) listed some of the vaccines they produce to immunize cattle FMD in Botswana. Aftosa (2014) stated that for a vaccine to be effective on a disease, it must contain the virus strain of the disease for which the vaccination is being done. However, since there are further subclasses of viruses within the seven strains of FMD, vaccines may not be capable of covering all of them. According to Thomson (1995), in Botswana, most vaccines are produced from the African buffalo. The author carried on stating that the SAT virus types continuously mutate, thus developing resistance to the vaccines. Therefore there is a requirement to always update the vaccines by studying the new virus strains resulting from mutation. This mutation is said to be a function of geographical location and conditions in the area. This means a specific vaccine must be produced for every FMD vaccination zone depending on the virus strain. However, it is difficult to study such mutations and continuously match the vaccines with the virus in the field. Also, because of the mutation of the virus strains, the cattle must be vaccinated with updated vaccines to give them immunity. This increases the cost of producing the vaccines and carrying out the operations.

2.6.3   Disinfection

Depa et al. (2012) outlined some chemicals used to kill the FMD virus as disinfection. They mentioned the use of household bleach mostly at a concentration of 3% to disinfect premises that are suspected of having been infected. However, the authors stated the disadvantage of this chemical as not being suitable for equipment and footpaths. Vinegar was also listed as a disinfectant when used at a concentration level of 4–5%. They also mentioned lye as a disinfectant when at 2% concentration. However, lye is said to be caustic. The authors also mentioned another chemical named Virkos S as one of the disinfectants and even praised it for being able to kill a wide variety of germs. Lastly, Depa et al. (2012) outlined peroxyacetic acid, which is said to have been approved by the Environment Protection Agency (EPA) for working against FMD viruses. In Botswana, disinfection is a widely used method of FMD control at veterinary checkpoints, as shown in Fig. 2.4. When leaving an area known for FMD, people are supposed to walk over a matt treated with a disinfectant.

Fig. 2.4 Veterinary checkpoints

Extra shoes are also supposed to be made to step on the matt. Vehicles must pass through a ditch filled with a disinfectant to kill a probable virus sitting on the wheels. This method of FMD control, as practiced in Botswana, leaves some blind spots for the spread of the FMD virus as humans can also carry the virus on clothes and skin. So there is no disinfecting of the FMD virus on human clothes and skin in Botswana. At times people even leave their extra shoes inside vehicles, which mean those shoes which may be carrying the FMD virus pass without being disinfected. Also, cars may not take the virus only on wheels but also on other parts.

2.6.4   Fencing

Fences are erected to control the movement of animals from one zone to another, as shown in Fig. 2.5. This is also to prevent wild animals from reaching perimeters which are meant for livestock rearing and vice versa. Such wild animals like buffalos may bring FMD into the cattle region. Botswana is divided into FMD zones which are separated by the veterinary cordon fences. Mogotsi et al. (2016) mentioned that a veterinary cordon fence exists along the Botswana-Zimbabwe border to control the movement of FMD vectors between the two countries. The authors outlined the disadvantages of cordon fences. The authors stated that damages usually occur to the cordon fences due to reasons such as criminal activities and wild animal damage. When an outbreak of FMD occurs, all products of FMD vulnerable animals are not allowed to cross from one zone to another. These products are mostly meat and milk. The veterinary officers are

Fig. 2.5   Erected fences to control movement of animals

working at the veterinary checkpoints search vehicles and luggage looking for the products. If the products are found, they are seized from the culprits and later disposed of in a manner such as incineration to eliminate any possible FMD virus within them. However, some people work hard in evading the confiscation of these products. This evasion includes cutting the fence to cross into the FMD-free zone with the products. Livestock theft is also a significant challenge prevailing in Botswana. Livestock thieves destroy barriers to allow their stolen livestock to cross. These criminal activities of smuggling increase the spread of FMD in two ways thus far. This is by the transfer of the products and livestock which may be carrying the FMD virus from one zone to another. The second identified way is by leaving a gap on the fence, which will allow more FMD carriers to cross the fence. In Botswana wild animals being elephants are also a challenge to the veterinary cordon fences as they always damage the fences. Along the Botswana-Zimbabwe border, there is a high density of elephants. The elephants mostly migrate to Botswana, destroying the fence in the process. The migration is also found in the borders on Botswana with Namibia and Zambia due to poaching in those countries.

2.6.5   Culling

This is the elimination of infected cattle through killing. The carcasses are then incinerated mostly in a pit to reduce the virus spread. However, the culling method may result in uninfected animals being killed, leading to unnecessary losses. Davies (2002) stated that culling could be combined with other methods such as vaccination. In Botswana, farmers are usually compensated after such an operation has been carried out. The compensations are generally a fraction of the total losses, which means the farmers suffer heavy losses. The government also loses financially due to settlements.

2.7   Impacts of FMD on Botswana's Economy

According to Mogotsi et al. (2016), the beef portion of Botswana's agricultural sector constitutes 80% out of the whole farming industry. FMD at times reduces this contribution by the farming industry. Thomson (1995) mentioned that as FMD arises, Botswana experiences problems of not being able to export beef to its most important markets. This then cripples the contribution to Gross Domestic Product (GDP) from beef exports. Pattnaik et al. (2012) stated that control measures are put in place to combat FMD. In Botswana, such methods as the erection of tight fences and preventive methods like vaccination lead to huge expenditures by the government. Some more economic impacts of FMD were outlined by Pattnaik et al. (2012). The authors mentioned that FMD leads to a reduced production of meat, milk, and wool, which are agricultural products that have a contribution to the economy. Agricultural drought power is also said to be negatively affected by FMD due to mortalities of animals. FMD typically results in the death of calves and abortions by some pregnant animals leading to a reduction in animal production. This production of animals is also affected in the sense that FMD leads to the creation of low-quality semen in bulls. The higher the population of animals, the more the animal-based agricultural products can come out. So a reduction in the community of animals will lead to reduced animal products. The authors continue to add on by stating that since the seven FMD virus strains have serotypes within them, it is difficult at times to control FMD resulting from such undefined strains which complicate the control of FMD to a more significant extent leading to more costs by employing improved methods of FMD control. FMD is also said to have threats on food security worldwide. This is because, upon FMD outbreak cases, animal food products such as meat and milk are given restricted movement, including cancellation of exports. This negatively impacts food security in those countries that import animal food products. The retail of animal products is also ceased leading to reduced sources of food for consumers and no income for retailers and producers.

2.8   Method of FMD Diagnosis

The results of the literature review have stated that a cow suffering from FMD will develop fever mainly as an increase in its body temperature. In particular, Business Queensland (2017) has provided a value of body temperature of 42 °C as the highest one. A suffering cow will also develop blisters on the tongue, which makes it hard for the cow to feed. Blisters also develop at the hooves resulting in a lame cow or having problems in moving from one place to another. To accurately detect FMD, the system uses the combination of knowledge of animal behavior and the sensor values to provide conclusions about FMD cases. For a given herd of cattle carrying the sensor nodes, the system first studies the cattle in the grazing land or anywhere they are kept. The movements of the cows are observed, that is, when they are grazing while not suffering from the disease. This monitoring process will take measurements of body temperature as well as the food intake of the cattle. After the typical observation has been done, a few cows are sampled from the large herd where the average values are determined. These average values of body temperature, movement, and feeding rate are then taken as reference values for the FMD diagnosis of the herd in question. A deviation from these observations should bring the farmers to attend. If the temperature stays above 40 °C with little movement and feeding rate for a prolonged period simultaneously, then a conclusion is made that a cow has caught FMD. The extended period being referred to here is taken as 2 h.

2.9 Sensor Nodes

The body temperature is measured through the use of the LM35 sensor. Detecting movement of jaws is chosen to recognize food intake. A sensor referred to as the piezoelectric thin-film sensor is mounted on the jaws to detect the chewing. For sensing the movement of the cow, an accelerometer is used. The chosen accelerometer is referred to as the ADXL335. Adding to these FMD parameters, the location of the cow must be known at all times. Knowledge of place is essential because if a cow is found to possess FMD, it must be located as soon as possible so that it can be transferred into quarantine. This is a safety measure to prevent the spread of the disease. For this reason, a Global Positioning System (GPS) module is added to the sensors to provide the location. The GPS module is therefore referred to as a position sensor in this document. The four sensors are interfaced to a PIC16F887 microcontroller which reads all the values provided by the sensors. Data read by the microcontroller is telemetered to the farmer using ZigBee. Therefore the sensor node is an assembly of four sensors, a microcontroller, and a wireless transmitter. A unique wearable item is designed, and the sensor node incorporated into it. The wearable item is placed around the head and neck area. Figure 2.6 is a block diagram of the sensor node, and Fig. 2.7 is a circuit diagram of the node.

2.9.1 ADXL335 Accelerometer

An accelerometer, as shown in Fig. 2.8, is used to calculate the rate of change of velocity of a moving or vibrating objecting as defined by Omega Engineering (2018). It uses a piezoelectric material whose volume changes by force due to the motion. The ADXL335 is an accelerometer that comes from Analog Devices (2009). The device provides signals which are conditioned and measures acceleration in three axes. It has a tunable bandwidth depending on the applications. The accelerometer

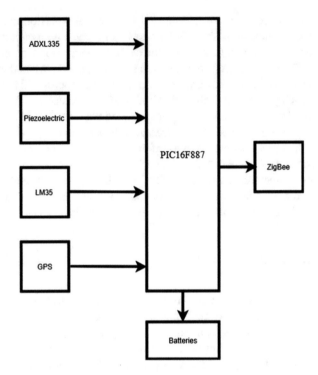

Fig. 2.6 Block diagram of the sensor node

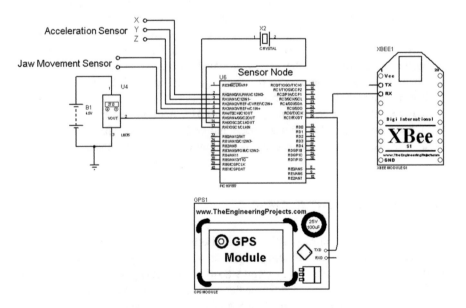

Fig. 2.7 Circuit diagram of the sensor node

Fig. 2.8 The ADXL335 accelerometer

Fig. 2.9 LM35 temperature sensor

includes an output for acceleration in each axis. The X and Y axes have a bandwidth range of 0.5–1600 Hz while the Z-axis has a bandwidth range of 0.5–550 Hz. From Analog Devices, it is available as a 4 mm × 4 mm × 1.45 mm with the packaging of type LFCSP_LQ. Its power supply ranges from 1.8 to 3.6 V and has an operating temperature range of −40 to +85 °C.

2.9.2 LM35 Temperature Sensor

An LM35 temperature sensor, as shown in Fig. 2.9, is calibrated to produce a voltage equivalent of 10 mV to every change of 1 °C in temperature; that is, it has a sensitivity of 10 mV/°C. The LM35 does not require any signal conditioning. Its output is directly fed into the microcontroller. This device has a temperature range of 0–150 °C.

2.10 Gateway

The Gateway (GW) is designed to be powered in three ways. This is from the AC mains power supply, solar panels, and batteries. The type of power supply to be used will depend on the positioning of the GW. If the GW is used inside a house, it will mostly be powered from the wall. If it is mounted outside, it will be powered by the

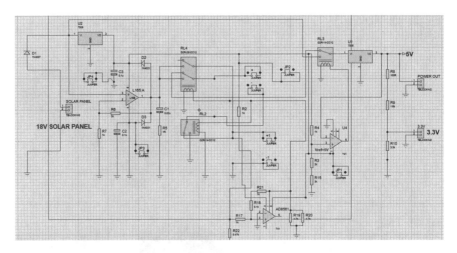

Fig. 2.10 Solar power and battery recharging circuit for the gateway

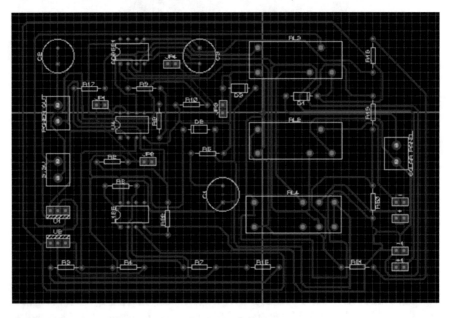

Fig. 2.11 PCB layout of the power supply circuit of the gateway

solar panels during the day and by the rechargeable batteries at night. Figures 2.10 and 2.11 show solar power and battery recharging circuit and PCB Layout of the power supply circuit of the gateway, respectively.

Figure 2.12 is the block diagram of the gateway (GW). It also uses the PIC16F887 microcontroller as the processor. Data from the sensor nodes come in through the ZigBee transceiver. The transceiver then relays the incoming data to the microcontroller Universal Asynchronous Receiver Transmitter (UART) Receive (Rx) pin. The data is then to be transmitted by the MCU to the users. Three ways are used to

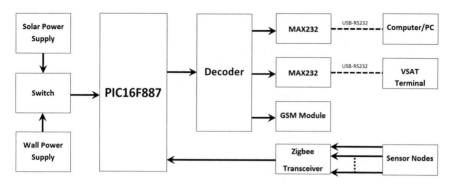

Fig. 2.12 Block diagram of the gateway

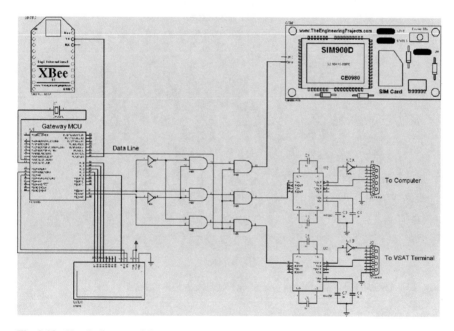

Fig. 2.13 Circuit diagram of the gateway

relay the data to users. These are a direct cable connection between the user's PC and the GW, VSAT terminal and a GSM module. However, the microcontroller under consideration has one Transmit (Tx) pin. A decoder is employed to send data to multiple devices through this pin. Figure 2.13 is the circuit diagram, while Fig. 2.14 is the whole FMD monitoring system.

2.11   Experiment

The project was tested in two forms. These were the simulation and hardware forms.

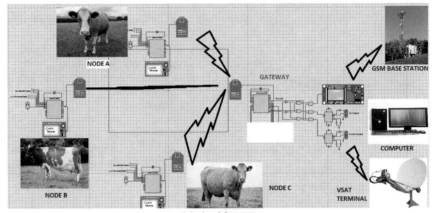

(a) Architecture

(b) Model

Fig. 2.14 Overall FMD monitoring system. (**a**) Architecture. (**b**) Model

2.11.1 Simulation

In a simulation, Proteus 8 Professional was used with MPLAB XC8 being used to write the program. Figure 2.15 showed an instance when the simulation in Proteus was running. There are two virtual terminals on the simulation. One is for displaying data as transmitted by the sensor node while the other is displaying data as received and routed by the gateway. The simulation used a star network topology, that is, when all the sensor nodes transmit directly to the portal with no hopping between the nodes.

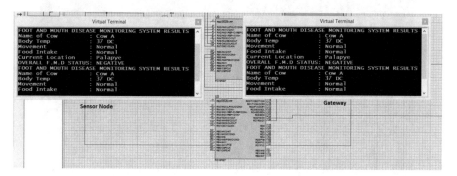

Fig. 2.15 Simulating the project in Proteus

2.11.2   Hardware

A circuit of the sensor node in Fig. 2.7 was built together with the ZigBee radio interfaced. A provision of signals from outside was used to replace the sensors. As proposed, the system is to use the accelerometer AXL335, a piezoelectric thin-film sensor, and the LM35 temperature sensor. All the three sensors require the use of Analog to Digital Converter (ADC) for reading the signal levels. The sensors since they were absent were replaced by the application of voltage signals at the ADC pins which would be interfaced to the sensors. These signals can be modulated just the way sensor outputs would vary with the varying stimuli. The applied signal levels are then read and interpreted as those coming from the real sensors. The test had variable resistors used for modulating the signals. A gateway circuit was also prototyped with a corresponding ZigBee radio put in place to receive the sensor node data. The gateway was then interfaced to a computer running LABVIEW software. The interface was obtained through the use of an RS232-USB cable with the RS232 end plugging to the Gateway and USB end going into the computer. LABVIEW in this case was used as a graphical user interface (GUI) of the system. Figure 2.16 is a picture of the prototype gateway. Figure 2.17 shows the sensor node, gateway, and computer put together. Figure 2.18 shows the LabVIEW interface with no results obtained yet. On the interface, there is a window labeled Foot and Mouth Disease Status. The purpose of this window is to display all data coming from the cattle being monitored. The displayed information is in the form of the text so that the user can easily interpret the results. On the right of the Foot and Mouth Disease Status window, there are three graphs which will plot the results of the three primary parameters being used to diagnose cattle for FMD. The graphs are for the plots of Food Intake, Body Temperature, and Speed. The instantaneous values of the settings are displayed on smaller windows near the top right corner of each graph. In the figure, all values are at 0 or minimum level since no data is coming in. The LabVIEW interface shows a window on which the communication port

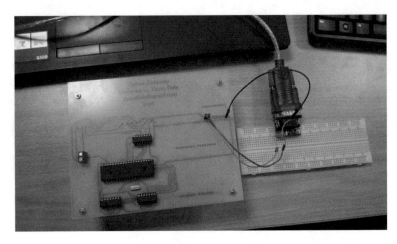

Fig. 2.16 The gateway circuit with RS232 cable plugged in

Fig. 2.17 The gateway and sensor node together with the data acquisition station

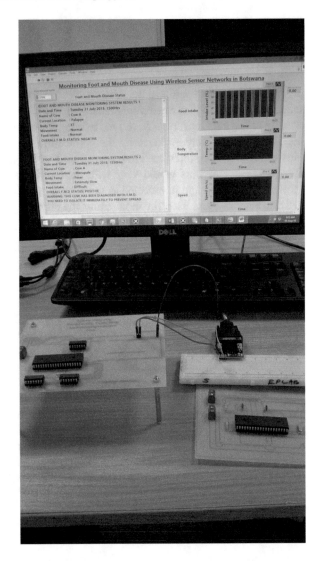

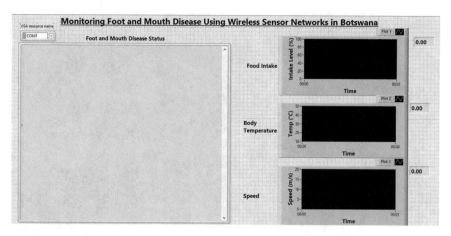

Fig. 2.18 LabVIEW graphical user interface of the FMD monitoring system

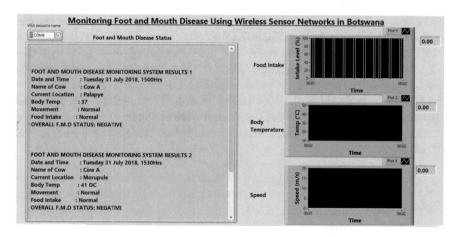

Fig. 2.19 LabVIEW showing the results of a negatively diagnosed cow

(COM PORT) of the serial data can be selected. The right port is selected from a dropdown list depending on the port at which the cable is plugged.

Figure 2.19 shows the LabVIEW interface displaying diagnosis results of a given cow. In this testing case, the sensor node was configured to transmit data to the base station every 30 min. The information arriving covers the Date and Time, Name of the Cow, location of the cow, the levels of the diagnosis parameters as well as the overall status of the cow in terms of FMD. The data in Fig. 2.19 is for a cow that is negatively diagnosed with FMD. The sensor node, depending on the parameter level, concludes whether the cow is negative or positive.

Figure 2.20 shows a case when the system detects FMD. When the cow is found to have been infected by the disease, a warning message is issued immediately to the user. In this case, the cow is seen to be suffering from fever, languid movement,

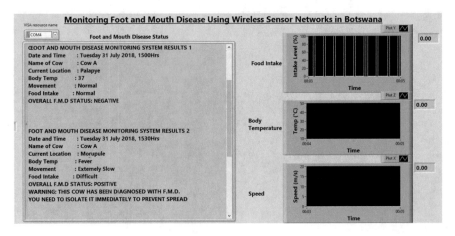

Fig. 2.20 LabVIEW GUI showing the results for a positively diagnosed cow

and difficulty in feeding. A combination of these is a clear sign of FMD as learned in the literature review section. Figure 2.20 displays a case when the system detects FMD in a 30 min time difference. At 1500 hours, according to the LabVIEW display, the cow was seen to be negative. Thirty minutes later, the signs and symptoms had reached the threshold levels indicating FMD. Such a case would occur due to the incubation period of FMD, which takes between 2 and 14 days. The signs and symptoms will gradually rise until the system detects them. If such an incident occurs, the user is advised to locate the cow and isolate it to prevent further spread of the disease.

References

Aftosa F (2014) Foot and Mouth Disease. [Online]. Available: http://www.cfsph.iastate.edu/Factsheets/pdfs/foot_and_mouth_disease.pdf

Analog Devices (2009) Small, Low Power, 3-Axis ±3 g Accelerometer. ADXL335 Datasheet, Norwood

Ashford DA (2019) [Online]. Available: http://www.msdvetmanual.com/generalized-conditions/foot-and-mouth-disease/overview-of-foot-and-mouth-disease

Botswana Vaccines Institute (n.d.) [Online]. Available: https://www.bvi.co.bw/content/id/27/Products/

Carbi (2019) [Online]. Available: https://www.cabi.org/isc/datasheet/82822

Davies G (2002) Foot and mouth disease. Res Vet Sci 73:195–199

Depa PM, Dimri U, Sharma MC, Tiwari R (2012) Update on epidemiology and control of foot and mouth disease—a menace to International trade and global animal Enterprise. Veterinary World 5(11):694–704

Derah N, Mokopasetso M (2005) The control of foot and mouth disease in Botswana and Zimbabwe. Tropicultura 23(special issue):1–7

Donaldson AI (1993) Epidemiology of foot-and-mouth disease: the current situation and new perspectives. In: Proceedings of an International Workshop Held at Lampang, Thailand, vol. 51, pp. 209, September 6–9

Jamal SM, Belsham GJ (2013) Foot-and-mouth disease: past, present and future. Vet Res 44(116):1–14

Mogotsi K, Kgosikoma OE, Lubinda KF (2016) Wildlife-livestock interface, veterinary cordon fence damage, lack of protection zones, livestock theft and owner apathy: complex socio-ecological dynamics in foot and mouth disease control in southern Africa. Pastoralism: Res Policy Pract 6:21

Omega Engineering (2018) Accelerometer: introduction to acceleration measurement. [Online]. Available: https://www.omega.com/prodinfo/accelerometers.html

Pattnaik B, Subramaniam S, Sanyal A, Mohapatra JK, Dash BB, Ranjan R, Rout M (2012) Foot-and-mouth disease: global status and future road map for control and prevention in India. Agric Res 1(2):132–147

Prinsloo T, de Villiers C, van Niekerk J. (2017) The role of the Namibian livestock traceability systems in containing the recent foot-and-mouth disease outbreak: case study from the northern parts of Namibia. In: 2017 1st International Conference on Next Generation Computing Applications (NextComp), Mauritius, pp. 30–35

Queensland Government (2017). Clinical signs of foot-and-mouth disease in cattle. Queensland Government, 24 May 2017. [Online]. Available: https://www.business.qld.gov.au/industries/service-industries-professionals/service-industries/veterinary-surgeons/foot-mouth-info/clinical-signs-cattle Accessed: 1 August 2018

The Cattle Site (2014). [Online]. Available: www.thecattlesite.com/diseaseinfo/243/footandmouth/

Thomson GR (1995) Overview of foot and mouth disease in southern Africa. Rev Sci Tech Off int Epiz 14(3):503–520

Van Tho L (2017). [Online]. Available: http://www.biopharmachemie.com/technical/health-management-disease-prevention-and-treatment-in-cattle-goat-sheep-horse/prevention-and-treatment-of-foot-and-mouth-disease-fmd.html

Chapter 3
Climate: Environmental Monitoring Using Wireless Sensor Network System

Abstract This chapter gives an overview of environmental monitoring systems using wireless sensor network, big data, and Internet of Things (IoTs). This chapter outlines the effect of climate change on wild animals and also discusses innovations in response to climate change. Electric fences are commonly used to control and manage the movement of animals in game reserves, private game and farms to restrict intruders such as unwanted predators and humans from the bound area. Wireless sensor network-based system for intruder detection and monitoring is presented in this system to minimize the human-animal disputes in Africa. It is challenging to observe elephants monitoring since these huge animals travel for very long distances. The biggest challenge in the existing wireless-based anti-poaching system is the limited network or no coverage. As a result, the non-monitored animals are simply subjected to poaching. Taking advantage of WSN, an anti-poaching system is proposed in this chapter.

3.1 Introduction

Hsu (2010) developed a real-time wireless sensor network (WSN) interface of a small-scaled wind-power electricity generator based on ZigBee and Bluetooth technologies. The author designed the self-supply power technique for WSN system to competently supply the generated power into the lead-acid battery.

Feng et al. (2010) proposed a model to decrease energy consumption based on an adaptive geographical fidelity (GAF) topology management protocol. Authors considered nodes in the vehicular ad hoc networks (VANETs) and their transmission range adjustment. It has been found from the study that the proposed genetic algorithm model saved 21.1% in comparison to the equal-grid model, while the adjustable-grid model saved 19.3% in comparison to the equal grid model.

Shanmuganathan et al. (2008) studied technologies to monitor and evaluate the climate change with weather and atmospheric environments and described the function of WSN that could be shared with the internet and employed in on-farm processes. Authors investigated three application scenarios and proposed a system for the relative study of viticulture management data from Chile and New Zealand. Authors investigated real-time sensors for the collection of data remotely employing related hardware and software infrastructure. Authors suggested that the database

© Springer Nature Switzerland AG 2020

A. Yahya, *Emerging Technologies in Agriculture, Livestock, and Climate*,
https://doi.org/10.1007/978-3-030-33487-1_3

could be used for developing a model to observe and study the results of climate variation on grapevine growth and wine standard.

Yang et al. (2013) proposed WSN-based online solution to monitor geological CO_2 storage and leakage. However, the proposed design could not accomplish low power consumption and, as a result, the device could not be a mobile one. Point Six™ (n.d.; http://www.pointsix.com/PDFs/3008-40-V6.pdf) utilized an NDIR sensor coupled with Wi-Fi to monitor the carbon dioxide level, but without humidity, atmospheric pressure, and light evaluation ability. EnOcean Alliance developed a self-powered monitoring system based on the Cozir® sensor (2013; https://www.envirotech-online.com/news/air-monitoring/6/gas-sensing-solutions/self-powered-co2-sensor-moves-into-volume-production/23884). However, the policy limited to a lesser number of measurement rates, amidst 9 and 2 per hour, subject to light intensity.

Donno et al. (2014) presented a radio frequency identification (RFID) augmented module for smart environmental sensing (RAMSES) to investigate RFID applications. To harvest RF energy, authors designed an RF-dc rectifier enhanced by a dc-dc voltage in silicon-on-insulator technology. The initial prototype not only outperformed in a laboratory but in the real world as well. Authors claimed and found from the demonstration that RF energy could be harvested up to 10 m of distance using RAMSES, which is the longest distance for any passive RFID sensors according to their study.

Taking into consideration the systems above, Folea and Mois (2015) presented a compressed battery-powered method that monitors the CO_2 level, temperature, relative humidity, absolute pressure, and intensity of light in indoor spaces.

It has been found from the experiments that the proposed scheme is not only cost-effective but also can operate on a single 3 V battery uninterrupted for up to 3 years without maintenance. Authors recommended that the proposed method can be employed in a full extent of monitoring applications as a module, or a cyberphysical, in the IoT or a WSN system.

Prabhu et al. (2014) presented a large-scale and long-term CO_2 monitoring system for greenhouses based on WSN. Authors recommended that sensors must be spread uniformly all over the greenhouse employing distributed clustering to control and monitor the environmental factors.

Jung (2015) investigated that Republic of Korea (ROK) selected 27 climate technologies to increase the greening of current businesses and to improve new green trades to support an ecological climate technology growth policy. The author pointed that government of Korea will invest more in R&D in the core climate technology programs, i.e., solar cells, fuel cells, bioenergy, rechargeable battery technology, and information technology (IT), by 2020. The author studied that implementation of climate technology will be more in effect in developing countries due its cost-effective applications.

Lee and Lin (2016) proposed an open-source wireless mesh network (WMN) module for environmental monitoring applications. Authors estimated and compared the proposed WMN module with ZigBee and found the proposed model outperformed ZigBee in terms of performance, consistency, and network complexity.

In a 20 node experiment setup the average package delivery ratio of the proposed model and ZigBee were 94.09% and 91.19% while 5.14% and 10.25% were observed for standard deviation, respectively. Authors emphasized on further investigation to enhance the performance and power consumption capability of the proposed model.

Mois et al. (2016) presented a cyber-physical method that monitors the environmental parameters remotely in indoor spaces employing IoT platform based on the existing IEEE 802.11 structure. It has been found from the results that the proposed system visualized and analyzed the data from any device linked to internet irrespective of location. Authors found that the communication protocol and the design of the nodes in the proposed system reduced the power consumption and increased battery life.

Mois et al. (2017) proposed IoT-based results for environmental monitoring using three different wireless sensors such as User Datagram Protocol (UDP), Hypertext Transfer Protocol (HTTP), and Bluetooth. Authors found that UDP packets consumed less energy as compared to ZigBee, at the cost of decreased transmission reliability. On the other hand, HTTP requests increased transmission reliability at the expense of battery life. Moreover, the authors investigated that BLE advertisement is appropriate for adding power generating elements and producing environmental data map for short packets.

Ferentinos et al. (2017) developed a WSN prototype to examine the outcomes of real greenhouse state on the understanding consistency of the sensor network measurements. Authors placed WSN prototype in a commercial greenhouse to study the spatial variation of the prevailing environmental circumstances. Considerable spatial variability in temperature and humidity has been observed with average diversity up to 3.3° C and 9% relative humidity and transpiration.

Schulz et al. (2018) presented energy-aware demand-side management (DSM) method to control manufacturing arrangements on the component level employing an IoT structure to minimize the appearing enduring power that must be well adjusted with grid-supplied power.

3.2 Microclimatic

Global warming, climatic extremes, severe drought, and flood have significantly influenced the ecosystem. Ma et al. (2015) examined the impacts of thrilling hydroclimatic distinctions in southeastern Australia on phenology and vegetation yield employing Imaging Spectroradiometer, Enhanced Vegetation, and Standardized Precipitation-Evapotranspiration Indices. Authors studied that hydroclimatic deviations on biotic and abiotic features are therefore of acute significance to precisely forecast the effects of future climate variation on ecosystem function and suggested a model that reflects fluctuating hydroclimatic sensitivities among biomes.

Burgess et al. (2010) developed a WSN-based system appropriate to plant physiological ecology. Authors found the proposed method showed adaptability about sensing operations that could be carried out and had a radio range that was appropriate for thorough monitoring, however not for broader-scale experimental designs. Authors found mobile phone network more suitable to transmit data and developed an approximately entire record of data at the office location.

Arx et al. (2012) studied long-term below-canopy microclimate to the local climate in an extensive range of Central Europe and other forest ecosystems in moderate scopes. It has been established from results that a reasonable overall outcome of the canopy on below-canopy microclimate with a reduction of daily highest air temperature and increase of daily minimum relative humidity was up to 5.1 °C and 12.4%, correspondingly.

The approaching of utilizing a thermometer screen to monitor forest microclimates is through in situ observations. However, the limitations are cost and labor force to control the vast region continuously. Moreover, practically it is not possible to arrange and regulate the sensors in complex terrain (Jin et al. 2018; Keller et al. 2008).

Jin et al. (2018) studied the applications of wireless sensors in forest microclimate monitoring system and the related observation precision, error causes, and correction methods. Authors studied and examined two typical subtropical forest ecosystems in Zhejiang Province, China, using WSN coupled with temperature and humidity sensors based on GreenOrbs deployments. Authors analyzed the experimental data and found variation in the observed errors due to the season, daily periodicity, and climate states. Authors recommend that shading measures must be taken into account for wireless sensors in the open air and to avoid plastic in the shell of the sensor.

3.3 Air Pollution Monitoring

Wireless sensor networks (WSN) offer an outstanding and economical solution to be employed in real-time automatic monitoring and control for air quality reasons. For instance, Völgyesi et al. (2008) designed the Mobile Air Quality Monitoring Network (MAQUMON) system coupled with gas sensors fixed on vehicles. The proposed system collected the data from the car and transferred to the servers using WiFi hotspot only to process and available on the Sensor Map portal. Hu et al. (2009) developed GSM short messages based Vehicular Wireless Sensor Network (VSN) system to monitor air quality. North et al. (2008) developed a mobile environmental sensing system to organize transportation and urban air quality. Authors set up sensor nodes on moveable vehicles to monitor traffic, weather, and pollutant concentrations and transmitted data into a platform that maintained mutually near real-time event management as well as more extended period planned outcomes. Jelicic et al. (2011) presented an Indoor Air Quality monitoring (IAQ) system coupled with a sensor network that incorporated a power management method to cut sensors energy consumption with an adaptive duty cycling methodology for metal oxide semiconductor (MOX) gas sensors.

Mansour et al. (2014) introduced a manageable energy-efficient air quality monitoring WSN-based system coupled with ozone, CO, and NO_2 gas sensors and Libelium's Waspmote. The proposed system utilized ZigBee for communication and Clustering Protocol for Air Sensor network (CPAS) to broadcast data over a short distance using multiple base stations.

Feng et al. (2010) proposed a model to decrease energy consumption based on an adaptive geographical fidelity (GAF) topology management protocol. Authors considered nodes in the vehicular ad hoc networks (VANETs) and their transmission range adjustment. The observation showed that the proposed genetic algorithm model saved 21.1% in comparison to the equal-grid model, while the adjustable-grid model saved 19.3% in comparison to the equal grid model.

Air quality monitoring offers raw quantities of gases and pollutants concentrations, which can then be examined and explained. Air pollution is an apprehension in many urban areas and is the principal purpose for respiratory difficulties among people; monitoring the air quality may comfort many travails from respiratory difficulties and diseases, and from then on advising engineering and policy decision-makers to reduce the pollution and provide healthy environment (Tapashetti et al. 2016). Tapashetti et al. (2016) designed a low-cost indoor air monitoring system to measure the concentration of carbon monoxide-CO and formaldehyde- HCHO gases. The proposed system communicated over cloud whenever the gases reached their threshold. Authors suggested that the enhanced version of the prototype could be used to monitor the high-fidelity radiations to investigate the outcomes of anthropogenic and environmental elements on intra-hour air standard.

Saha et al. (2017) proposed a system to determine the pollutants in nature employing the IoT. The UVI-01 sensor is an ultraviolet light sensor. In the proposed order, the collected data from the sensors will be analyzed over the cloud, and the actionable report will be produced against pollution.

Arroyo et al. (2018) designed cloud-based a cost-effective, small size, and low consumption wireless gas sensor network for environmental applications and air quality detection. The insight volume of the system was tested with benzene, toluene, ethylbenzene, and xylene (BTEX) and to process the data, pattern recognition and artificial intelligence techniques were used. A significant enhancement in the signals was perceived from 80% (PV1) to 100% (PV4).

Miralavy et al. (2019) proposed gas emission monitoring system based on WSN employing MAC layer protocols. Authors monitored the pollution level, emitted by the vehicles, and recommended that the data transmission interval must be higher than 0.1 s to prevent redundant information to be transmitted more than once in a day. Authors recommended that the sensor node should be installed in the Environmental Control Unit (ECU) and mounted on the vehicle to avoid modification in the car system.

3.4   Sensor Nodes Deployment

Chakrabarty et al. (2002) proposed grid coverage planning for efficient inspection and target position in distributed sensor networks. Authors used an integer linear programming (ILP) to reduce the amount of sensors placement whereas confirming broad coverage of the sensor field. Authors studied various kinds of sensors and found that the installation of sensors with more extensive detection ranges will cost more. However, the proposed model affected the complexity due to the nonlinear formulation and noncircular recognition part in some applications.

Meguerdichian and Potkonjak (2003) suggested an ILP formulation of coverage based on the optimization issue. However, their proposed ILP model considered only the 1-Coverage based on a single sensor.

The optimum position and placement of sensors reduce the cost in WSN design. Boubrima et al. (2015) proposed two real-time monitoring and integer linear programming (ILP) models that guarantee pollution coverage and network connectivity together. Authors matched the two models concerning implementation time and recommended the second flow-based design.

3.5   Outdoor Air Pollution Monitoring

Kwon et al. (2007) developed an air pollution monitoring system coupled with dust, temperature- relative humidity and CO_2 sensors based on ZigBee platform for the ubiquitous city. In the proposed method, TOSSIM simulator was used for analyzing three kinds of routing protocols based on ARM7 architecture. Authors recommend flooding routing protocol to cover a wide area with low cost. Jung et al. (2008) proposed an air pollution monitoring scheme based context model and flexible data acquisition policy to observe the position of air pollution in a distant place. In the recommended method, the authors placed 10 routers together with 24 sensors to collect environmental data. Authors suggested the heterogeneous geosensor data abstraction and mixture for an advanced environment.

Hasenfratz et al. (2012) designed and carried out a low-power and low-cost android-based GasMobile system to monitor participatory air pollution. Authors suggested and recommended to use GasMobile to develop combined high-resolution air pollution maps to analyze and track the air pollution thoroughly.

Yu et al. (2012) proposed a novel air pollution monitoring scheme deployed on a public transportation system employing Chemical Reaction Optimization (CRO) to achieve better optimization. Authors suggested using the different sensor for changed routes to cut the sensing break and enhance the precision. Kasar et al. (2013) presented air pollution monitoring to monitor CO_2, NO_2, and SO_2 gases from the environment using WSN based on ZigBee module. Authors employed the air quality index (AQI) to calculate the level of health apprehension for a precise location using instinctive colors scheme. Kadri et al. (2013) presented a real-time air quality monitoring system comprised of several distributed monitoring stations using machine-to-machine communication. Authors arranged four static solar-powered gas sensor nodes over an area of 1 km². Authors used the website and mobile application to deliver that data collected over 4 months to end-users.

Navarro et al. (2013) installed a long-term WSN-based environmental monitoring system at the Audubon Society of Western Pennsylvania (ASWP), USA. Authors deployed 42 sensor nodes to monitor the environment for the past 2 years based on TinyOS platforms and XMesh routing protocol software. However, the physical nodes in the proposed system failed due to overuse of routing path across the network. Resultantly, the packet was retransmitted and lost in ASWP. Authors recommended using the online management system to produce real-time network performance. Ustad et al. (2014) proposed an air pollution monitoring method coupled with Mobile Data-Acquisition Unit (MobileDAQ) and a fixed Internet-Enabled Pollution Monitoring Server-based ZigBee module. The Mobile-DAQ unit incorporates a single-chip microcontroller, air pollution sensors array, and GPS Module. Public transports were used to collect pollutant gases from sensors and transmitted to a central node for online access with the help of the Google Maps interface.

Oikonomou et al. (2014) introduced WSN, coupled with an array of eight polymer-coated capacitive sensors and low-power read-out electronics, to detect low concentrations of specific volatile organic compounds (VOCs) available at industrial installations. Authors recommended that the system can be employed to monitor the industrial environment in real time.

Climate Change and Wild Animals
Unstable weather patterns in Southern Africa have silent yet very implicative threats to wild animals. Due to harsh weather conditions and drought, the survival rate of wild animals, particularly young ones, is rapidly decreasing. South Africa's Cape Town is well known for its natural fynbos vegetation, which covers the region's splendid mountains, valleys, and coastal plains. Currently, Cape Town is experiencing prolonged drought periods that have since led to water shortages and a large number of fynbos species in threat of extinction.

In Africa, the diaspora and migration regions for wild animals are now being occupied by humans due to rapid human population expansions. Wild animals move to these areas for survival purposes, but they end up being killed by humans in a war of occupation. Figures 3.1, 3.2, and 3.3 show the dry season in Botswana.

3.6 Elephants

The African elephant is the world's biggest walking species as shown in Fig. 3.4. Elephants can be readily acknowledged by their trunks apart from their big size, and these trunks develop throughout the lifetime of the elephant. Elephants need 150–300 L of water a day to survive, to cool down on their bodies to drink, bathe, and spray water because they are susceptible to high temperatures. African elephant

Fig. 3.1 Central Kalahari Game Reserve (CKGR), Botswana [Photography by Abid Yahya]

Fig. 3.2 Central Kalahari Game Reserve (CKGR), Botswana [Photography by Abid Yahya]

Fig. 3.3 Nata Bird Sanctuary, Botswana [Photography by Abid Yahya]

Fig. 3.4 Elephants at Chobe National Park, Botswana [Photography by Abid Yahya]

Fig. 3.5 Hyena at Mashatu Game Reserve, Botswana [Photography by Abid Yahya]

breeding is also linked to rainfall in the sense that these elephants' birth rates are proportional to the peaks of rains. The worst drought has been reported in Gonarezhou National Park, Zimbabwe, where 1500 African elephants died in the period from 1991 to 1992.

3.7   Spotted Hyenas

Spotted hyenas, one of Africa's top predators as shown in Fig. 3.5, also known by their scientific name Crocuta crocuta, have good sense of smell and sharp eyesight and hearing capabilities up to 3 mi. These scavengers live jointly in a group and are guided by female leaders. It has been proved from research that the environment plays a crucial role in regulating the hyena immune defense system [https://www.nationalgeographic.com/animals/mammals/s/spotted-hyena/].

3.8   Rhino

The population of rhinos as shown in Fig. 3.6 in the Republic of Botswana is currently facing a severe threat due to poaching since rhinos use watering holes daily; they are an easy target. Rhino horn black market price is about $65,000 per kilo, which is worth more than gold.

Fig. 3.6 Rhinos at Khama Rhino Sanctuary, Botswana [Photography by Abid Yahya]

Khama Rhino Sanctuary is a conservation park, situated around 25 km north of Serowe, to protect the vanishing white and black rhinos [https://www.cadip.org/protecting-the-endangered-rhinos-of-botswana]. The conservation park has a varied range of animals together with kudus, impalas, gemsboks, giraffes, and a diversity of other animals and birds. It acts as one of a few recognizable ways to conserve rhino's from extinction in Botswana. Black rhino males merely create reproduction areas wherever surface water is everlasting. The water-reliant white rhino requires everyday mud baths for thermoregulation and parasite control. Two species of rhino have been extinguished in the last century, the Western Black Rhino in Cameroon and the Indochinese Javan Rhino in Vietnam. Two additional species—Northern White Rhino and Sumatran Rhino's continental population are approaching extinction and are recorded on the IUCN Red List as being critically endangered [https://www.iucn.org/content/rhinos-rise-africa-northern-white-nears-extinction].

3.9 Lion

The number of African lions has continued to drop in the last two decades as shown in Fig. 3.7. This is due to multiple life-threatening factors, one being climate change as it severely disrupts their reproductive cycles.

There was a massive die-off in the era 1994–2004 due to the outbreak of a disease called distemper. Researchers found that severe drought was the main reason for the epidemic. Rain showers can persuade changes in habitat fitness, which is capable of adjustment between predator and prey relations.

Fig. 3.7 Lion at Central Kalahari Game Reserve (CKGR), Botswana [Photography by Abid Yahya]

Fig. 3.8 Wild dog at Moremi Game Reserve, Botswana [Photography by Abid Yahya]

3.10   African Wild Dogs

African wild dogs, as shown in Fig. 3.8, were initially known for their adaptation to the heat. However, new studies reveal that it is becoming too hot for this species to hunt and the number of puppies that live is dropping.

Fig. 3.9 Hippo at Chobe River, Botswana [Photography by Abid Yahya]

Increase in drought severity and frequency causes mass mortality in herbivores as well. Climate change for wild dogs implies hotter days, shorter periods of hunting, and less food. In Botswana, the average amount of wild dog puppies who touched the date of their birth dropped by 35 percent from 5.1 per litter between 1989 and 2000 to 3.3 between 2001 and 2012, with temperatures increasing by 1.1C during the similar era. [http://www.reportfrompeter.com/index.php/tag/extinction/page/2/].

3.11   Hippo

During dry seasons, hippos spend more than 12 h in lakes. Due to increasing temperatures, the depletion of natural water habitats provides the species with various difficulties. Overcrowding enables quicker disease transmission, while infighting over shrinking regions has resulted in a male loss. Figure 3.9 shows hippos enjoying in Chobe River, Botswana.

3.12   Cheetah

Because of inbreeding in fragmented populations, the cheetah already suffers from an absence of genetic diversity—reproduction rates are reduced, and the few surviving descendants are more susceptible to illness. The African cheetah is the fastest animal in the world, but in the face of climate change, it is racing against its near-threatened status. The prey populations of the cheetahs are decreasing in some fields, and the cheetahs have had to alter their diets as a consequence. An increase in temperatures has had an impact on the reproductive capacity of this big cat. Male cheetahs showed reduced testosterone concentrations and nearly ten times reduced sperm counts than your cat in the house.

3.13   Giraffe

The giraffe is the highest on earth animal, as shown in Fig. 3.10. These beautiful creatures are one of Africa's most famous species. They are renowned for their long necks, long legs, and brown and white patterns covering their bodies. Loss of habitat has dramatically decreased giraffe populations once widespread across the continent. Although they can also resist prolonged dry seasons, ecological boundaries such as rivers or human-made constructions may impede their capacity to disperse as climate changes. Since the 1980s, the amount of giraffes has fallen by as much as 40%, a recent study on threatened animals suggests.

Fig. 3.10 Giraffe at Mashatu Game Reserve, Botswana [Photography by Abid Yahya]

3.14   Birds

Birds migrate and arrive sooner on their nesting grounds, and the nesting grounds to which they move are not as far away as they used to be, and in some nations, the birds are no longer even leaving, as the climate is appropriate throughout the year. Birds were laying eggs sooner than usual in the year.

Over 30% of breeding birds are already decreasing and need conservation intervention. In the future, the impact of climate change on birds will become more severe unless greenhouse gas emissions are reduced, and birds need to adapt to change to safeguard natural resources. Figures 3.11 and 3.12 show lilac-breasted roller and blue waxbill at Mashatu and Moremi Game Reserve, Botswana, respectively.

Innovations in Response to Climate Change
It is an increasing need in Africa to promote sustainable engineering through the implementation of environmental management systems (EMS). Over one-third of the continent's countries receive the greater part of their energy from hydroelectric plants, and droughts have rendered that supply unreliable over the previous few years. Countries mainly relying on fossil fuels were disturbed by price volatility and increased laws. The price of renewable technology has dropped dramatically at the same moment. And scientists find that the continent has more potential for solar and wind energy than earlier thought.

Fig. 3.11 Lilac-breasted roller at Mashatu Game Reserve, Botswana [Photography by Abid Yahya]

Fig. 3.12 Blue waxbill at Moremi Game Reserve, Botswana [Photography by Abid Yahya]

3.15   Windmills

The energy policies in Africa until recently focused on fossil fuels and hydroelectricity. In contrary, developed countries have been focusing on green generation methods such as wind power and solar power stations. Wind parks and solar fields are being exploited for energy extraction from wind and solar in developed countries. Objectives have been set for green energy to total energy ratio with that ratio timed to increase annually for countries such as America, the United Kingdom, and South Africa. On- and offshore wind parks are being built annually as wind generation has been tested and proved not damaging to the environment.

A windmill converts wind energy to usable energy through the rotation of a wheel made up of variable blades. It is an environmentally friendly way of pumping water which has been used for a long time in Africa. It does not require human power and uses wind as a renewable source of energy. The unavailability of safe drinking water characterizes most rural communities in Africa. Examples of wind energy harvesting in Botswana are investment by Wind Edge Botswana in a wind energy project to produce electricity in Botswana's Kweneng District. The average wind speed at the project site is at least 6 m/s at 80 m above ground level. These wind speeds are sufficient to drive modern, 50–100 m high wind turbines. The project will use 50× 2 MW of big wind turbines with a capability of 100 MW installed. The project will generate 210 GWh of electricity per year with an assumed capacity factor of 0.24. The electricity produced by the turbines will be marketed into the national grid and electricity from the coal thermal power stations in Botswana will be displaced, thereby decreasing GHG emissions in the form of CO_2 in particular.

Fig. 3.13 Windmill, Ghanzi, Botswana [Photography by Abid Yahya]

The expected life span of the turbine is 20 years, but with appropriate maintenance, this can be expanded. Figure 3.13 shows a windmill in Ghanzi, Botswana.

3.16 Solar Panels

Solar panels are generally installed on rooftops or in large areas to gather as much sunlight as possible without blockade. However, researchers are now looking into harvesting solar energy from raindrops, solar panel roads, windows, and cell phone screens. At the Central Kalahari Game Reserve (CKGR) in Botswana, some water pans are made by installing solar panels around a designated area, as shown in Figs. 3.14 and 3.15.

Wireless Sensor Network-Based System for Intruder Detection and Monitoring

A passive infrared sensor (PIR) is used to sense motion; it is mostly used to sense whether a human/animal has moved in or out of the sensors range. The small sizes of PIR sensors single them out as low-power, cost-effective, and user-friendly sensors.

The upper electrode of a pyroelectric material is covered with a black layer. The temperature of the absorbent layer increases once the infrared radiation hits the layer, and resultantly charges the surface. The fluctuation in temperature is significant to attain better SNR, and also, the thickness of the chip maintains the heat capacity. To reach an SNR of 1, the noise equivalent power (NEP) is frequently used to set against dissimilar since it defines the radiant power that is imposed on the detector (https://www.infratec.de/).

Fig. 3.14 Solar Panel at Central Kalahari Game Reserve (CKGR), Botswana [Photography by Abid Yahya]

Fig. 3.15 Bore Hole at Central Kalahari Game Reserve (CKGR), Botswana [Photography by Abid Yahya]

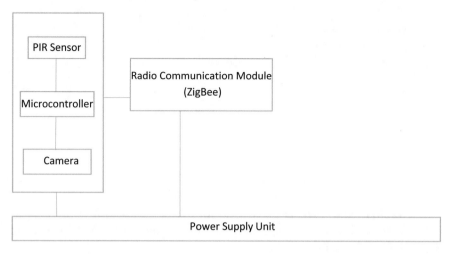

Fig. 3.16 Sensor node block diagram

$$\text{NEP} = \frac{N}{R_v}$$

where N is the noise density and R_v represents responsivity.

PIR sensing is determined by the direction of movement and the space of the body from a sensor (Zappi et al. 2010).

Moghavvemi and Seng (2004) installed PIR sensors on the wall to spot and discover the movement of a human in a protected room. Lee et al. (2006) presented a smart home-based indoor location-aware system using an array of PIR sensors. The authors positioned a sensor on a ceiling to detect the location of residents in a room.

Luo et al. (2009) proposed a PIR sensor-based system that is attached to a ceiling of an observed area. The proposed system deployed four sensor units, each comprised of five sensor detectors to inspect the position of a person in motion.

Electric fences are commonly used to control and manage the movement of animals in game reserves, private game and farms to restrict intruders such as unwanted predators and humans from the bound area. At times, some of the big animals become rowdy around the fence, which often results in electrocution (de Bruin 2017). Human-animal disputes are high in Africa, and resources are damaged due to the struggle between them.

Sensor nodes are installed remotely to monitor animals in a farm at various points of the farm boundaries. Hence, it is essential to have a system architecture that will optimize the lifetime and self-sufficiency of sensor nodes. This reduces the requirement for frequent component replacement and eventually reduces overall maintenance costs.

In this work, all the sensor nodes communicate directly with the base station to avoid the complex routing.

Sensor nodes consist of a PIR sensor, Arduino microcontroller, ArduCam shield with OV2640 camera module, ZigBee module, and power supply unit, as shown in Fig. 3.16.

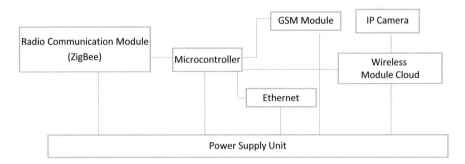

Fig. 3.17 Base station/gateway architecture

The central gateway comprises Arduino microcontroller, ZigBee, GSM module for SMS and GPRS, WiFi Module, Ethernet, IP camera for surveillance and power supply unit as shown in Fig. 3.17.

The system is powered with a battery together with a solar system to enhance the operational lifetime.

The PIR sensor detects movement and triggers the ArduCam to capture a picture. The ArduCam is relatively competent as it decreases the complexity of camera control connections, incorporates image sensor OV2640, offers small size, and has user-friendly hardware connections. At the transmitting and receiving ends, an SD card shield works as a data storage unit before wireless transmission. It transmits the captured images to an Arduino board through SPI. The captured image is sent to a gateway using a secure channel through a ZigBee module. The gateway then assembles the data in a local database for processing. A secure cloud connection is established with the gateway for image classification and visualization processes. The captured image is checked against a set of sample images that are already stored in the cloud. If the intruder classifies as a threat, an alert notification in the form of an SMS and an intruder image is sent to a farmer.

Along with this alert notification, there is a blinding flash of light with an irritating noise that is used to chase the intruder. A Bluetooth solar-powered speaker coupled with a flashlight is deployed at the boundaries of the farm. In this project the IP camera is used as an additional monitoring tool; it can be used in instances where the user would like to verify the source of the intrusion or if the image sent to the user is not clear. The user can access the IP camera to get a live feed through a smartphone or any other gadget that supports the application. Figure 3.18 shows overall WSN-based system architecture.

Anti-Poaching System Using Wireless Sensors
An anti-poaching system is a group of mechanisms and devices composed of an interconnecting wireless sensor network to stop poaching by monitoring and collecting data at a central point.

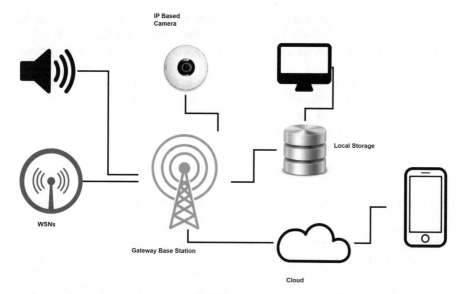

Fig. 3.18 Overall WSN-based system architecture

3.17   Problem Analysis

Anderson (2014) investigated that poachers have fully prepared strategies and plans for identifying, killing, and dehorning targeted animals.

It is challenging to observe elephants monitoring since these huge animals travel for very long distances. The biggest challenge in the existing wireless-based anti-poaching system is the limited network or no coverage. As a result, the non-monitored animals are simply subjected to poaching.

Mulero-Pázmány et al. (2014) recommended drones technologies to use as anti-poaching tactics. Authors explained three different modes of drone operation such as manually, first-person view mode, and radio control conventional mode.

Koen (2017) studied DNA mapping approaches to determine the accurate where-abouts of a dead animal. The author discussed that in 2002, Singapore officials carried out the DNA mapping system to find out 6.5 tons of ivory in a container to Zambia, with the help of microsatellite.

Banzi (2014) investigated mobile biological sensors (MBS)-based anti-poaching system, deployed in Tanzania National Parks. The author studied that the system gathered and proceeded with the collected data to find out if there is an unexpected movement of animals. Nevertheless, there is an option that the variation might be detected after an incident has happened.

Sudha et al. (2018) presented a WSN-based wildlife monitoring system. The proposed system tracks the animal and records the health condition of the animal and updates the concerned authority remotely.

Wall et al. (2014) presented four algorithms to monitor the African elephant movement using a cloud-based system. The authors suggested that the data collected from the activities of an animal can support the researcher's experiments and design.

Kumar and Hancke (2015) proposed a ZigBee-based real-time system to monitor temperature, heart rate, and physiological condition of the animal.

Nobrega et al. (2018) proposed IoT based on an animal behavior monitoring system. Authors stated that the system comprises an IoT local network to collect animal's information and direct ovine within winery independently.

3.18 The Methodology of the Anti-Poaching System

The project aims to detect the temperature of the target if the temperature does not fall within the desired range; the data will be transmitted to the concerned authority.

There are several functions of the organism, such as nutrition, reproduction, activity, stress response, and health, that are related to temperature. This is the main reason why it is essential to study the temperature parameter. However, the research is not limited to only one parameter, in future parameters such as heart monitoring can be looked into in further details as it is among essential features related to functions of the organism. Variations of these functions can be monitored by monitoring of animal's body temperature (Sellier et al. 2014). One existing technology to monitor the body's temperature with direct contact is through the use of thermistor or thermocouples; this addresses the question of what is needed to measure temperature.

After the temperature has been detected by the sensor, before it is transmitted to the user, it will need to be processed so that it will be able to be transmitted through the wireless channel. In this project, Arduino microcontroller is used to process the data, as shown in Fig. 3.19.

Fig. 3.19 Arduino

3.18.1 Arduino Technology

Arduino is an open-source hardware and software technology based on single-board microcontroller kits (Hari Sudhan et al. 2015).

A microcontroller is programmed with using Arduino programming language, which is close as C and C++ programming languages. The Arduino programming language uses an Arduino integrated development environment (IDE) for writing the code that will control the microcontroller.

Sensors and transmitting components are interfaced to the Arduino; the transmitter is responsible for conveying the data to the destination.

3.18.2 433 MHz ASK Transmitter and Receiver

The RF Transmitter unit consists of a SAW resonator tuned at 433.XX MHz operation frequency, a switching transistor, and some passive modules, as shown in Fig. 3.20 (Fahmida et al. 2016).

Data can be displayed either using mobile phones or personal computer; however, in this project, 1602 Liquid Crystal Display is used as shown in Fig. 3.21. The overall system block diagram is shown in Fig. 3.22.

3.18.3 Interfacing Temperature Sensor, RF Transmitter, and Receiver to Arduino

Temperature sensor TMP36GZ consists of three pins of which pin 1 is voltage supply between 2.7 V and 5.5 V, pin 2 output voltage from the temperature sensor, and pin 3 the ground pin.

Fig. 3.20 433 MHz ASK transmitter and receiver

Fig. 3.21 Liquid crystal
display

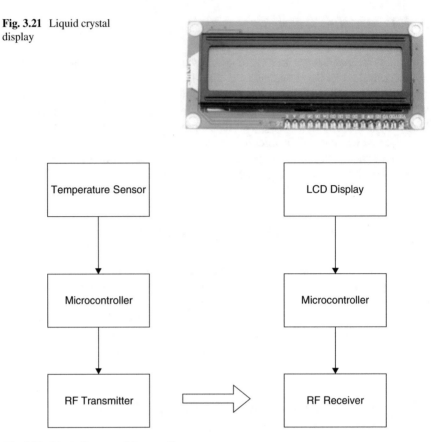

Fig. 3.22 Block diagram of the overall system

The temperature sensor output pin is connected to an analog input pin of the Arduino. Figure 3.23 shows the RF transmitter being interfaced to Arduino, coupled with the temperature sensor. Figure 3.24 shows the receiver and LCD interfaced to Arduino.

After the temperature is read from the sensor, the value is interpreted as a voltage signal by Arduino. The analog voltage reading will range from 0 to 1023, 0 being no voltage and 1023 being 5 V. The following calculations are the steps to convert the voltage value to degrees Celsius.

$$\text{Temp} = \text{voltagereading} / 1024$$
$$\text{Temp} = \text{Temp} \times 5$$
$$\text{Temp} = \text{Temp} - 5$$
$$\text{Temp} = \text{Temp} \times 100$$

After this conversion, the value is converted to string since it was initially declared as double for it to be able to be transmitted over a wireless channel. Through the SPI library, the data will be sent to the channel using the RF ASK transmitter. Figure 3.25 shows the flowchart of data flow from the sensor to the LCD.

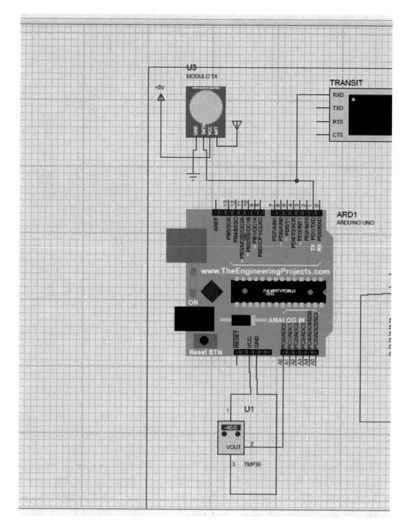

Fig. 3.23 Transmitter and temperature interfaced to Arduino

3.19   Simulations and Hardware Setup of Anti-Poaching System

Proteus software simulation was used to test the working of the components and program. Figure 3.26 shows all the components being connected to the Proteus software simulation. As can be seen from the simulation, the LCD is displaying the temperature being detected by the temperature sensor at the transmitter side.

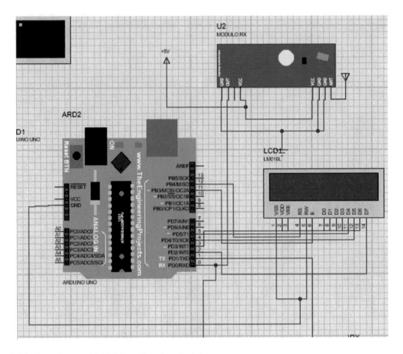

Fig. 3.24 Receiver and LCD interfaced to Arduino

The connection is followed just as from interfacing temperature sensor with Arduino, interfacing transmitter and receiver with Arduino, and lastly interfacing LCD with Arduino, as it can be seen from the simulation. Figure 3.27 shows the hardware components being interfaced. The temperature sensor is interfaced to Arduino board, and the RF ASK transmitter is interfaced to Arduino board. Figure 3.28 shows the receiver components hardware, the RF ASK transmitter being interfaced to Arduino, and the LCD is interfaced to Arduino. As can be seen, the main aim was to be able to view the temperature being read by the temperature sensor. As can be seen, the LCD shows the temperature being detected by the temperature sensor. Figure 3.29 shows the model to emulate kind of real-life situation.

In the future, the Global positioning system will be incorporated into the system to enhance the tracking of endangered species. Also, the system in the future can be able to be deployed in such a way that it can be able to work in multispecies. Even in future, the simulation will be done on several simulation tools; the prototype will have a power supply circuit that can regulate power into the circuit to 5 V and using a solar system for powering the system. Moreover, the installation of cameras will also be considered as a future development for this project.

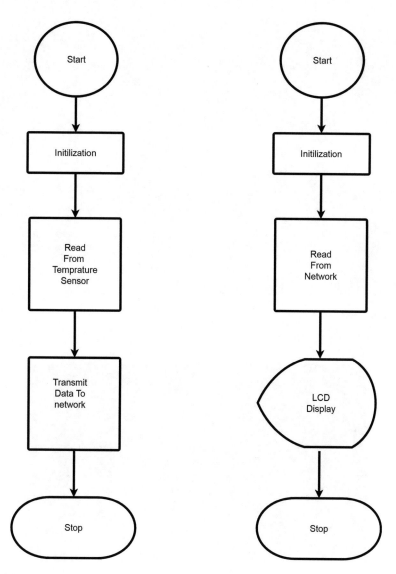

Fig. 3.25   Flowchart of data flow from the sensor to the LCD

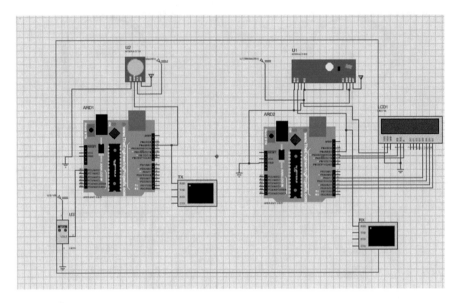

Fig. 3.26 Complete system

Fig. 3.27 Transmitter in hardware

Fig. 3.28 Receiver in hardware

Fig. 3.29 Prototype model
of anti-poaching system

References

Anderson AJ (2014) Modern intelligence measures to combat animal poaching: a conservation & counterterrorism strategy. Master Thesis, Institute for Intelligence, Mercyhurst. Edu, May 2014
Arroyo P, Lozano J, Suárez JI (2018) Evolution of wireless sensor network for air quality measurements. Electronics 7(342):1–16
von Arx G, Dobbertin M, Rebetez M (2012) Spatio-temporal effects of forest canopy on understory microclimate in a long-term experiment in Switzerland. Agric For Meteorol 166–167:144–155

Boselin Prabhu SR, Sophia S, Balamurugan P (2014) Environmental monitoring and green-house control by distributed wireless sensor network. Scholars Journal of Engineering and Technology (SJET) 2(4A):511–516

Boubrima A, Matigot F, Bechkit W, Rivano H, Ruas A (2015). Optimal deployment of wireless sensor networks for air pollution monitoring. 24th International Conference on Computer Communication and Networks (ICCCN), Las Vegas, NV, pp. 1–7

de Bruin L (2017) Electric fences and small animals—a deadly combination. https://www.up.ac.za/research-matters/news/post_2477106-electric-fences-and-small-animals-a-deadly-combination

Burgess SSO, Kranz ML, Turner NE, Cardell-Oliver R, Dawson TE (2010) Harnessing wireless sensor technologies to advance forest ecology and agricultural research. Agric For Meteorol 150(1):30–37

Chakrabarty K, Iyengar SS, Qi H, Cho E (2002) Grid coverage for surveillance and target location in distributed sensor networks. IEEE Trans Comput 51(12):1448–1453

De Donno D, Catarinucci L, Tarricone L (2014) RAMSES: RFID augmented module for smart environmental sensing. IEEE Trans Instrum Meas 63(7):1701–1708

Enocean Alliance (n.d.) Self-powered CO2 sensor moves into volume production. [Online]. Available: http://www.enocean-alliance.org/en/gss-seamless-sensing-co2-sensor-moves-into-volume-production/

Environmental Technology (2013) Self-powered CO2 sensor moves into volume production. [Online]. Available: https://www.envirotech-online.com/news/air-monitoring/6/gas-sensing-solutions/self-powered-co2-sensor-moves-into-volume-production/23884

Fahmida A, Alim SMA, Islam MS, Kawshik KBR, Islam S (2016) 433 MHz (Wireless RF) communication between two Arduino UNO. American Journal of Engineering Research (AJER) 5(10):358–362

Feng W, Alshaer H, Elmirghani JMH (2010) Green information and communication technology: energy efficiency in a motorway model. IET Commun 4(7):850–860

Ferentinos KP, Katsoulas N, Tzounis A, Bartzanas T, Kittas C (2017) Wireless sensor networks for greenhouse climate and plant condition assessment. Biosyst Eng 153:70–81

Folea SC, Mois G (2015) A low-power wireless sensor for online ambient monitoring. IEEE Sensors J 15(2):742–749

Hari Sudhan R, Ganesh Kumar M, Udhaya Prakash A, Anu Roopa Devi S, Sathiya P (2015) Arduino atmega-328 microcontroller. International Journal of Innovative Research in Electrical, Electronics, Instrumentation and Control Engineering 3(4):27–29

Hasenfratz D, Saukh O, Sturzenegger S, and Thiele L (2012) Participatory air pollution monitoring using smartphones. 2nd International Workshop on Mobile Sensing, April 16–20, China, ACM 978-1-4503-1227-1/12/04

Hsu C-L (2010) Constructing transmitting interface of running parameters of small-scaled wind-power electricity generator with WSN modules. Expert Syst Appl 37(5):3893–3909

Hu S-C, Wang Y-C, Huangl C-Y, Tseng Y-C (2009) A vehicular wireless sensor network for CO2 monitoring. IEEE Sensors 2009 Conference, Christchurch, New Zealand, pp. 1498–1501

Jelicic V, Magno M, Paci G, Brunelli D, Benini L (2011) Design, characterization and manage-ment of a wireless sensor network for smart gas monitoring. 4th IEEE International Workshop on Advances in Sensors and Interfaces (IWASI), 2011, pp. 115–120, 28–29 June

JF Banzi A (2014) Sensor based anti-poaching system in Tanzania. Int J Sci Res Publ 4(4):1–7

Jin J, Wang Y, Jiang H, Chen X (2018) Evaluation of microclimatic detection by a wireless sensor network in forest ecosystems. Scientific Reports, 8, Article number: 16433

Jung TY (2015) Climate technology promotion in the Republic of Korea. Adv Clim Chang Res 6(3–4):229–233

Jung YJ, Lee YK, Lee DG, Ryu KH, Nittel S (2008) Air pollution monitoring system based on geosensor network. IGARRS, 2008, IEEE International Geoscience and Remote Sensing Symposium, Boston, MA, pp. III-1370–III-1373

Kadri A, Yaacoub E, Mushtaha M, Abu-Dayya A (2013) Wireless sensor network for real-time air pollution monitoring. IEEE 1st International Conference on Communications, Signal Processing, and Their Applications (ICCSPA), Sharjah, pp. 1–5.

Kasar AR, Khemnar DS, Tembhurnikar NP (2013) WSN based air pollution monitoring system. International Journal of Science and Engineering Applications (IJSEA) 2(4):55–59

Keller M, Schimel DS, Hargrove WW, Hoffman FMA (2008) Continental strategy for the National Ecological Observatory Network. Front Ecol Environ 6:282–284

Koen H. Predictive policing in an endangered species context: combating rhino poaching in the Kruger National Park. Ph.D. thesis, 2017

Kumar A, Hancke GP (2015) A ZigBee-based animal health monitoring system. IEEE Sensors J 15(1):610–617

Kwon J-W, Park Y-M, Koo S-J, Kim H (2007) Design of air pollution monitoring system using ZigBee networks for ubiquitous city. IEEE 2007 International Conference on Convergence Information Technology (ICCIT 2007), Gyeongju, pp. 1024–1031, 2007

Lee H, Lin H (2016) Design and evaluation of an open-source wireless mesh networking module for environmental monitoring. IEEE Sensors J 16(7):2162–2171

Lee S, Ha KN, Lee KC (2006) A pyroelectric infrared sensor-based indoor location-aware system for the smart home. IEEE Trans Consum Electron 52(4):1311–1317

Luo X, Shen B, Guo X, Luo G, Wang G (2009) Human tracking using ceiling Pyroelectric Infrared sensors. 2009 IEEE International Conference on Control and Automation, Christchurch, pp. 1716–1721

Ma X, Huete AR, Moran S, Ponce-Campos GE, Eamus D (2015) Abrupt shifts in phenology and vegetation productivity under climate extremes. Ecosystem Functional Response to Drought Jgr: Biogeosciences 120(10):2036–2052

Mansour S, Nasser N, Karim L, Ali A (2014) Wireless sensor network-based air quality monitoring system. 2014 International conference on computing, networking and communications (ICNC), Honolulu, HI, pp. 545–550

Meguerdichian S, Potkonjak M (2003) Low power 0/1 coverage and scheduling techniques in sensor networks.

Miralavy SP and Atani RE and Khoshrouz N. (2019) A wireless sensor network based approach to monitor and control air pollution in large urban areas. 4th Conference on Contemporary Issues in Computer Information and Sciences, 23–25 Jan. 2019, Kharazmi University, Karaj, Iran

Moghavvemi M, Seng LC (2004) Pyroelectric infrared sensor for intruder detection. IEEE Region 10 Conference TENCON, Chiang Mai, 2004, Vol. 4, pp. 656–659

Mois G, Sanislav T, Folea SC (2016) A cyber-physical system for environmental monitoring. IEEE Trans Instrum Meas 65(6):1463–1471

Mois G, Folea S, Sanislav T (2017) Analysis of three IoT-based wireless sensors for environmental monitoring. IEEE Trans Instrum Meas 66(8):2056–2064

Mulero-Pázmány M, Stolper R, van Essen LD, Negro JJ, Sassen T (2014) Remotely piloted aircraft systems as a rhinoceros anti-poaching tool in Africa. PLoS One 9(1):1–10

Navarro M, Davis TW, Liang Y, Liang X (2013) A study of long-term wsn deployment for environmental monitoring. IEEE 24th Annual International Symposium on Personal, Indoor, and Mobile Radio Communications (PIMRC), London, 2013, pp. 2093–2097

Nobrega L, Tavares A, Cardoso A, Goncalves P (2018) Animal monitoring based on IoT technologies. 2018 IoT Vertical and Topical Summit on Agriculture - Tuscany. IOT Tuscany 2018(May):1–5

North R, Richards M, Cohen J, Hoose N, Hassard, J, Polak, J (2008) A mobile environmental sensing system to manage transportation and urban air quality. IEEE International Symposium on Circuits and Systems, ISCAS 2008, pp. 1994–1997, May 2008

Oikonomou P, Botsialas A, Olziersky A, Stratakos I, Katsikas S, Dimas D, Sotiropoulos G, Goustouridis D, Raptis I, Sanopoulou M (2014) Wireless sensor network based on a chemocapacitive sensor array for the real-time monitoring of industrial pollutants. Procedia Engineering 87:564–567

PointSix. (n.d.) WiFi 2000 ppm CO2 and Temperature Transmitter 3008–40-V6. Point Six Wireless, Data Sheet. [Online]. Available: http://www.pointsix.com/PDFs/3008-40-V6.pdf

Saha HN et al. (2017) Pollution control using Internet of Things (IoT). 8th Annual Industrial Automation and Electromechanical Engineering Conference (IEMECON), Bangkok, pp. 65–68

Schulz J, Popp RS-H, Scharmer VM, Zaeh MF (2018) An IoT based approach for energy flexible control of production systems. Procedia CIRP 69:650–655

Sellier N, Guettier E, Staub C (2014) A review of methods to measure animal body temperature in precision farming. American Journal of Agricultural Science and Technology 2(2):74–99

Shanmuganthan S, Ghobakhlou A., Ghobakhlou A. (2008) Sensors for modeling the effects of climate change on grapevine growth and wine quality ICC08 Proceedings of the 12th WSEAS international conference on Circuits, Heraklion, Greece, pp. 315–320

Sudha BS, Yogitha C, Sushma KM, and Pooja Bhat (2018) Forest monitoring system using wireless sensor network. International Journal of Advances in Scientific Research and Engineering (IJASRE), 4(1):127–130

Tapashetti A, Vegiraju D, Ogunfunmi T (2016) Iot-enabled air quality monitoring device: a low-cost smart health solution. IEEE Global Humanitarian Technology Conference (GHTC), Seattle, WA, pp. 682–685

Ustad V, Mali AS, Kibile SS (2014) Zigbee based wireless air pollution monitoring System using low cost and energy efficient sensors. Int J Eng Trends Technol 10(9):456–460

Völgyesi P, Nádas A, Koutsoukos X, Lédeczi Á (2008) Air quality monitoring with SensorMap. In Proceedings of the 7th International Conference on Information Processing in Sensor Networks, pp. 529–530

Wall J, Wittemyer G, Klinkenberg B, Douglas-Hamilton I (2014) Novel opportunities for wildlife conservation and research with real-time monitoring. Ecol Appl 24(4):593–601

Yang H, Qin Y, Feng G, Ci H (2013) Online monitoring of geological CO_2 storage and leakage based on wireless sensor networks. IEEE Sensors J 13(2):556–562

Yu JJQ, Li VOK, Lam AYS (2012) Sensor deployment for air pollution monitoring using public transportation system. IEEE Congress on Evolutionary Computation, Brisbane, QLD, 2012, pp. 1–7

Zappi P, Farella E, Benini L (2010) Tracking motion direction and distance with Pyroelectric IR sensors. IEEE Sensors J 10:1486–1494

Index

© Springer Nature Switzerland AG 2020
A. Yahya, *Emerging Technologies in Agriculture, Livestock, and Climate*,
https://doi.org/10.1007/978-3-030-33487-1

Printed in the United States
By Bookmasters